THE BRAIN WITHOUT OXYGEN

The Brain Without Oxygen

Causes of Failure-Physiological and Molecular Mechanisms for Survival

by

Peter L. Lutz

Florida Atlantic University,
Boca Raton, U.S.A.

Göran E. Nilsson

University of Oslo,
Oslo, Norway

and

Howard M. Prentice

Florida Atlantic University,
Boca Raton, U.S.A.

KLUWER ACADEMIC PUBLISHERS

DORDRECHT / BOSTON / LONDON

A C.I.P. Catalogue record for this book is available from the Library of Congress.

ISBN 1-4020-1165-2

Published by Kluwer Academic Publishers,
P.O. Box 17, 3300 AA Dordrecht, The Netherlands.

Sold and distributed in North, Central and South America
by Kluwer Academic Publishers,
101 Philip Drive, Norwell, MA 02061, U.S.A.

In all other countries, sold and distributed
by Kluwer Academic Publishers,
P.O. Box 322, 3300 AH Dordrecht, The Netherlands.

The first two editions of this book were published
by **Landes Bioscience** and **Chapman & Hall**.

Printed on acid-free paper

Printed in the Netherlands.

PREFACE

This third edition was motivated from comments and feedback from our readers to provide an increased molecular perspective. Not only has the original material been updated but new sections have been added on the mechanisms of brain hypoxia tolerance shown by some exceptional vertebrate species. New chapters have been added on oxygen sensing, molecular aspects of brain ischemia and acclimation to brain hypoxia.

CONTENTS

CHAPTER 1

INTRODUCTION

The human brain is the most complex organ in the body being made up of around 10^{11} intercommunicating nerve cells that can be subdivided into many hundreds of different cell types. This complexity comes, however, at the cost of a correspondingly high vulnerability for failure – there are so many steps to go wrong. The brain is committed to a continual active state – unlike muscle it never shuts down – and it has an obligatory high level of energy consumption. Any halt in energy supply, such as caused by asphyxia (hypoxia), ischemia or cerebral hemorrhage will quickly result in irreversible damage. Indeed focal ischemia (stroke) is a major mortality factor, causing some 150,000 deaths per year in the USA, and many who survive will suffer from the debilitating and tragic effects of permanent focal brain damage.

However, the dependence of the brain on an uninterrupted supply of oxygen is not just a human phenomenon, it is, in fact, common to all vertebrates – including fish, reptiles and birds; indeed, comparatively few species can withstand severe hypoxia and almost none can survive chronic anoxia lasting more than minutes. There are however exceptions. The epaulette shark *(Hemiscyllium ocellatum)*, and common frog (*Rana sp*) can tolerate anoxia for several hours. A few species are truly anoxia tolerant in that they can survive anoxia from days to months. Among the fishes these include the crucian carp (*Carassius carassius*) and goldfish (*C. auratus*), and in the reptiles the freshwater turtles *Chrysemys picta* and *Trachemys scripta*. The crucian carp can live in anoxic water for months at temperatures close to freezing. Similarly the turtle, *Trachemys scripta*, can withstand anoxia for 48 h at 20 °C and at least 3 months at 3 °C.

These anoxia tolerant animals employ many extraordinary measures to cope with the internal changes that occur during prolonged anoxia. For example, in the goldfish and crucian carp the accumulating lactate is converted to ethanol and the anoxic turtle can accommodates plasma calcium concentrations as high as 50 mM, magnesium levels of 15 mM and lactate levels of 200 mM. However, as extraordinary as these whole body responses might be, without

The Brain Without Oxygen. Third Edition. Peter L. Lutz, Göran E. Nilsson and Howard M. Prentice.

doubt the mechanisms that allow their brains to survive anoxia are the key adaptations for anoxia tolerance.

The purpose of this book is to describe the physiological systems that promote brain survival in the absence of oxygen in anoxia tolerant species. We make comparisons with the processes that occur in the mammalian brain during the initial few min of asphyxiation, when anoxia is combated and the degenerative changes that occur during the subsequent anoxic failure. We anticipate that studies using anoxia tolerant species as models will help to identify potential targets for human clinical intervention. As will be shown, the most interesting are those that are involved in the suppression of metabolic and functional activities to new set points well below their normoxic minima or maintenance levels. This involves a drastic, but balanced, reduction in ATP demand and supply pathways. These include activating mechanisms which cause the arrest of specific ion membrane channels, suppress electrical activity and synaptic inhibition. Signal transduction pathways are activated that lead to the activation of some genes and the down regulation of others. Identifying the key processes that are protected during anoxia and discovering those that are allowed to lapse in the anoxia tolerant brain may provide a fresh insight into the survival hierarchy of the many systems that collapse in the energy deprived mammalian brain.

ACKNOWLEDGMENTS

The writing of this book was supported in part by the Swedish Council for Forestry and Agricultural Research, the Swedish Natural Science Research Council and Florida Atlantic University Foundation.

CHAPTER 2

NORMAL BRAIN FUNCTION

We start this book by briefly sketching out some basic aspects of brain function that are of particular importance for understanding the mechanisms of death and survival in anoxia.

1. NEURONAL COMMUNICATION

The nerve cell, or neuron, is a highly specialized cell type that has the ability to propagate signals, consisting of waves of depolarization moving along the neuronal axon, and transmit these signals to other neurons by releasing neurotransmitters at synapses – the sites of communication between neurons. The neuron receives signals from other neurons through receptor molecules – membrane bound proteins that have the ability to bind neurotransmitters with a high degree of affinity and specificity. When a neurotransmitter binds to a receptor, conformational changes in the receptor molecule will directly, or indirectly, activate (or deactivate) ion channels, another specialized group of membrane-bound proteins, thereby changing the ion conductance of the neuronal membrane

1.1 The Membrane Potential

Like other cells, the distribution of ions on either side of the nerve cell membrane is very uneven, with much more Na^+, Cl^- and Ca^{2+} in the extracellular fluid surrounding the cell than in the, and higher concentrations of K^+ inside than outside the cell (Table 2-1).

These concentration differences are maintained by specialized membrane bound proteins, ion pumps, that actively transport specific

---

The Brain Without Oxygen. Third Edition. Peter L. Lutz, Göran E. Nilsson and Howard M. Prentice.

ions against their electrochemical gradients. For Na^+ and K^+ an ATP consuming pump, the Na^+/K^+ ATPase (the Na^+/K^+ pump) transfers Na^+ out of and K^+ into the cell in the stoichiometric ratio 3 Na^+ ions and 2 K^+ ions for each ATP molecule consumed.

The lipid bi-layer of the nerve cell membrane is virtually impermeable to electrically charged molecules and ions can only move down their electrochemical gradients through water-filled protein pores called ion channels. For the purpose of the present discussion brain ion channels can be divided into 2 major categories:

The first category consists of channels that open or close in response to a change in the membrane potential. There are two types in this category – (1) those whose activity is markedly voltage dependent and are closed at the resting membrane potential, commonly called voltage dependent ion channels (Fig. 2.1); and (2) those that are weakly voltage dependent and are open at the resting membrane potential, usually referred to as leakage channels.[46]

Table 2.1 Intracellular and extracellular concentrations of ions in the vertebrate brain

Ion	Extracellular concentration (mM)	Intracellular concentration (mM)
Na^+	150	30
K^+	< 3	100
Ca^{2+}	1.3	0.0001
Cl^-	130	5

Values from Hansen.[33]

The second major category consists of channels that open when they are activated by specific transmitters, neuromodulators or intracellular messenger molecules. The ligand-gated channels are included in this category.

As will be discussed below, leakage K^+ channels contribute to the resting potential, voltage gated ion channels play a pivotal role in the formation and termination of action potentials, and ligand gated channels produce synaptic potentials.

Under resting conditions the neuronal membrane has a much greater permeability for K^+ than for Na^+, with the result that a potential difference, the *resting membrane potential,* is produced. This potential is close to the Nernst equilibrium for K^+, viz. about -70 mV (the inside being negatively charged compared to the outside of the neuron). Any change in the membrane potential in the direction of 0 mV is referred to as a depolarization, while a change towards a more negative membrane potential is referred to as a hyperpolarization.

At the resting membrane potential, the negative charge of the cell interior provides a strong driving force for Na^+ and Ca^{2+} to enter into the neuron. Also the concentration gradients for Na^+ and Ca^{2+} over the cell membrane provide forces for an inward flow of these ions. As seen in Table 2.1, Na^+ is about 5 times more concentrated in the extracellular fluid compared to the cytoplasm, and the chemical gradient for Ca^{2+} is enormous, the concentration difference over the cell membrane for Ca^{2+} being about 10,000 fold.

K^+ and Cl^- also have large chemical gradients across the neuronal membrane (Table 2.1), K^+ being approximately 33 times more concentrated in the cytoplasm and Cl^- about 26 times more concentrated in the extracellular fluid. However, because the electrochemical equilibrium for these ions are close to the resting membrane potential the driving forces for these ions to move down their concentration gradients are small or nonexistent.

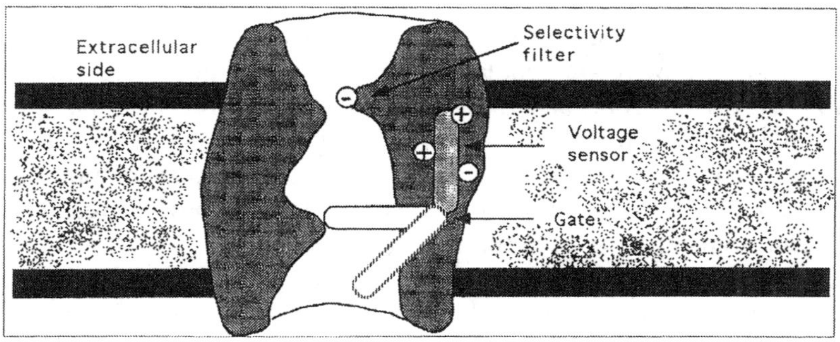

Fig. 2.1 The function of a voltage-gated ion channel according to the working hypothesis suggested by Hille.[35] *The selectivity filter is here assumed to be a narrow part of the channel where the size and charge of this pore determines what ion may pass. The voltage gate itself is guided by a voltage sensor. The charged groups on the voltage sensor cause it to move in response to the membrane potential.*

1.2 The Action Potential

An action potential is a large and rapid change of voltage across the cell membrane, caused by the opening and closing of voltage gated ion channels (Fig 2.2). It is initiated by a comparatively small depolarization of the membrane potential (typically 15 mV) which when it reaches about -55 mV causes a sudden opening of the voltage gated Na^+ channels. The result is a steep rise in the membrane permeability for Na^+ and a depolarizing shift towards the Na^+ equilibrium potential (about +55 mV). Within 1 ms the membrane potential reaches +20 to +40 mV. The action potential is terminated by the opening of voltage gated K^+ channels and the closing of the voltage gated Na^+ channels, which quickly act to repolarize the membrane. The voltage gated Na^+ channels close within a few milliseconds after their opening, a typical feature of voltage gated Na^+ channels. The K^+ channels open at a membrane potential of about -50 mV, but their activation is delayed until after the action potential has been fully developed.

By allowing Na^+ to flow into the cells, the open Na^+ channels produce a local depolarization of the membrane which in turn causes neighboring voltage gated Na^+ channels to open. The result is the rapid propagation a wave of depolarization, the action potential, along the axon of the neuron. When an action potential reaches the axonal terminals, where the neuron makes synaptic connections with other neurons, voltage gated Ca^{2+} channels in the presynaptic membrane are opened and the ensuing inflow of Ca^{2+} initiates the process responsible for neurotransmitter release (new or none).

Note that since the electrochemical potentials for K^+ and Cl^- (approximately -80 to -60 mV) are close to the resting membrane potential, increasing the membrane conductances to either K^+ or Cl^- by opening K^+ or Cl^- channels will tend to increase their influence on the resting potential, a “clamping” effect that will make it more difficult for the cell to depolarize.

1.3 Neurotransmitters and Neuromodulators

We now move on to the events at the synapse (Fig. 2.3 and 2.4), the site of neurotransmitter release and communication between neurons. Neurotransmitters, the chemical messengers between neurons, are ubiquitous regulators of neuronal activity and can exert either an excitatory or an inhibitory effect on a neuron. Excitatory neurotransmitters mediate a depolarization of the neuronal membrane by opening Na^+ and Ca^{2+} channels, thereby promoting action potentials

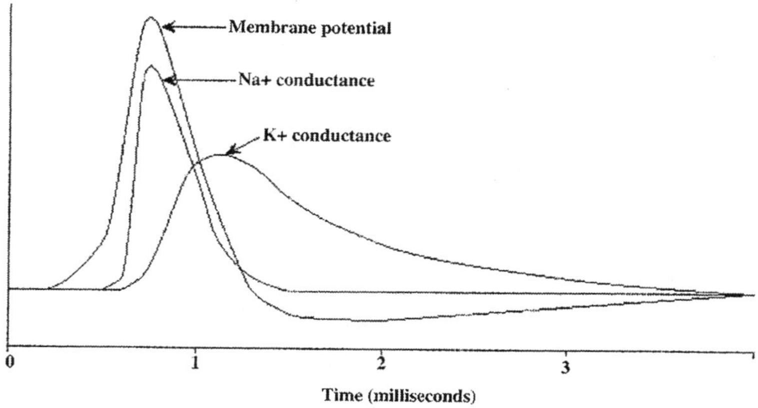

Fig. 2.2 The action potential. The changes in the membrane potential during an action potential are caused by the opening of voltage-gated Na^+ and K^- channels. The Na^+ channels are responsible for the development of the action potential while the delayed opening of the K^+ channels terminates the action potential.

and further transmitter release. Inhibitory neurotransmitters act by increasing the resistance for depolarization, usually by opening Cl^- or K^+ channels. As mentioned above the opening of Cl^- or K^+ channels will tend to clamp the membrane potential and prevent depolarization and the development of action potentials.

Ligand-gated ion channels are gated by neurotransmitters or other small molecules. This term usually only refers to those ion channels where the binding site for the ligand is situated on the protein complex that forms the channel. Such ionotropic receptor-channel complexes are responsible for the fast action of neurotransmitters, where it is a matter of microseconds between neurotransmitter release and the opening of the ion channels. However, there is also a large group of synaptic receptors that does not form integrated parts of ion channels. This group is responsible for slower synaptic responses that can persist for seconds or minutes. They act by coupling with GTP-binding (G) proteins, which in turn may stimulate or inhibit the production of intracellular second messenger molecules, and which finally activate the ion channels.

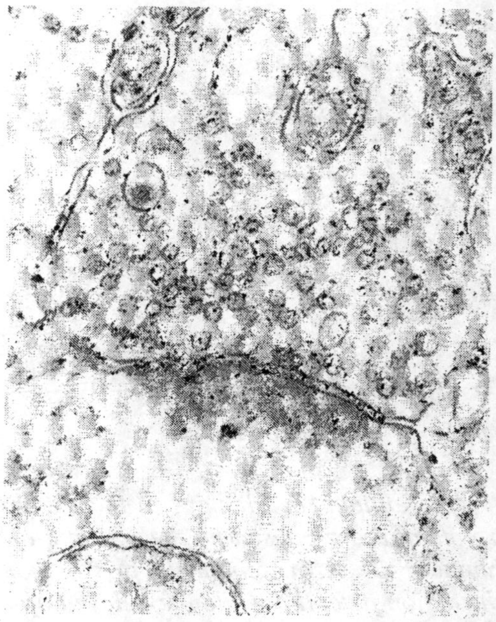

Fig. 2.3 A synapse in the vertebrate brain illustrated by a transmission electron micrograph from the rainbow trout telencephalon at x 70 000 (Hylland P, Johansson D and Nilsson GE). Note the presence of synaptic vesicles in the presynaptic compartment (above), the thin synaptic cleft, and the postsynaptic compartment (lower part of the picture). The dark shading of the postsynaptic membrane is likely to reflect the high density of receptors and connected molecules in this area.

A substance that can activate a certain receptor is said to be an *agonist* of that receptor, those that act by blocking receptors are termed *antagonists*. The neurotransmitters themselves are of course agonists of their own receptors. For many neurotransmitter receptors, synthetic agonists and antagonists have been produced and it is not an overstatement to say that these agents have played a fundamental role in the study of synaptic transmission and receptor function in the vertebrate brain. Moreover, synthetic receptor agonists or antagonists play a pivotal role in experiments aimed at understanding the mechanisms of anoxic brain damage as well as in attempts made to

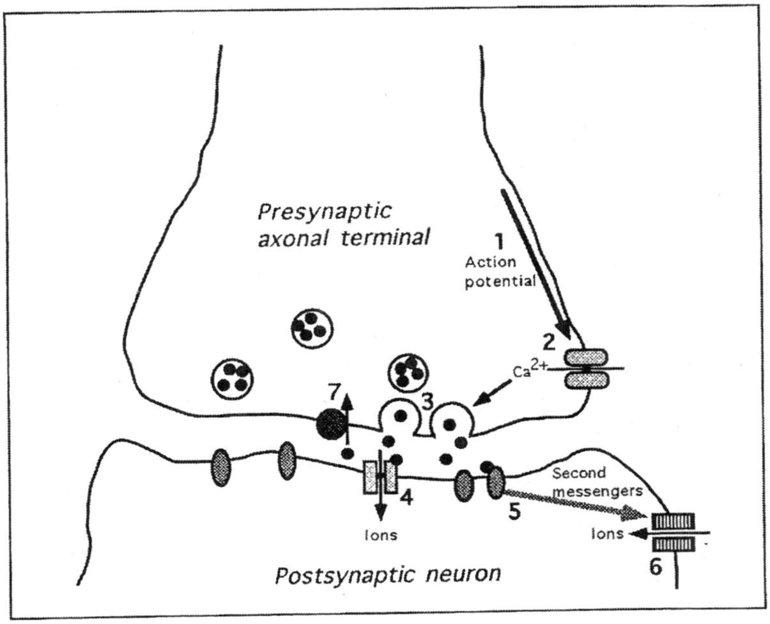

Fig. 2.4 A schematic representation of a synapse and the mechanisms involved in synaptic neurotransmission. When an action potential (1) reaches the presynaptic axonal terminal, voltage gated Ca^{2+} channels are activated (2). The inflow of Ca^{2+} triggers the synaptic vesicles to fuse with the presynaptic cell membrane, thereby releasing their content of neurotransmitter molecules (3) into the synaptic cleft. The neurotransmitters that have been released can either directly bind to and activate ligand-gated ion channels (4) or bind to receptor molecules (5) that initiates the activation of ion channels through a second messenger mechanism (6). The actions of released neurotransmitters are often terminated by uptake of the neurotransmitter into the presynaptic compartment (7) or into other adjacent cells.

minimize the deteriorating effects of anoxia. Neurotransmitters are released by exocytosis when the vesicles in which they are stored fuse with the presynaptic membrane at specific regions called docking sites.[55] This fusion is initiated by an increase in intracellular Ca^{2+} brought about by the opening of voltage gated Ca^{2+} channels in response to depolarization. The lag time between membrane depolarization and vesicle – membrane fusion is extremely short, less than a millisecond, the rapid response is due to many vesicles being already docked at the presynaptic membrane. Partly, the release of the neurotransmitters may be done through a "kiss-and-run" mechanism by which the synaptic vesicles undergo fast and efficient recycling: fusing with the neuronal membrane without intermixing of membranes [106]. Several proteins have been identified that appear to be involved in the process of neurotransmitter release, but the exact nature of the docking and fusing mechanisms still remains to be clarified.[55,106] Each neurotransmitter has its own set of receptors which it binds to after having been released. Neurotransmitter receptors are imbedded in the neuronal membranes, generally the postsynaptic membrane, but some receptor types are also situated presynaptically or on other cell types such as glial cells.

However, not all ion channel ligands are released from vesicles, which is one of the criteria for classical neurotransmitters. For example, adenosine, a compound of special interest in anoxia, is not stored in vesicles. Nevertheless, it has its own specific receptors with which it interacts when released into the extracellular space. Substances such as adenosine are often loosely defined as neuromodulators. Another example is provided by those K^+ channels that are regulated by intracellular ATP levels (K_{ATP} channels). These channels are of particular interest here since one of the early events in anoxic failure is a depletion of ATP. In this case the ligand, ATP, is interacting with its channel without being released to the extracellular space.

The most abundant neurotransmitters in the vertebrate central nervous system (CNS) are the amino acids glutamate, GABA (gamma-aminobutyric acid) and glycine. Glutamate acts as an excitatory neurotransmitter by binding to receptors that activate ion channels with conductances for Na^+ and Ca^{2+}, while GABA and glycine have inhibitory actions by binding to receptors coupled to Cl^- or K^+ channels. Some other neurotransmitters, such as the monoamines and acetylcholine, have more diversified actions – the same transmitter can produce either excitatory or inhibitory effects depending upon the receptor and ion channel it is interacting with.

1.3.1 Glutamate

Glutamate is the major excitatory neurotransmitter in the vertebrate brain and is by far the neurotransmitter that has attracted most attention with regard to anoxic/ischemic brain damage. Glutamatergic synapses are found in virtually every part of the central nervous system. Glutamate interacts with at least four main types of receptors of which three are ionotropic receptors (i.e. coupled directly to an ion channel) and at least one is a metabotropic receptor (i.e. exerting its effect through stimulation of intracellular messenger systems).

The metabotropic glutamate receptor (mGluR), of which there seem to be several subtypes, is G-protein coupled and its activation causes an increase in the intracellular levels of diacylglycerol, inositol 1,4,5-triphosphate (IP_3) or Ca^{2+}. The effects of metabotropic receptor activation is complex, and can include both an increase in neuronal excitability and a decrease in excitatory synaptic transmission[3]. The mGluR is pharmacologically characterized by being activated by quisqualate (QUIS) and it is therefore sometimes named the QUIS metabotropic receptor. However, to avoid confusion, it should be pointed out that QUIS also activates an ionotropic glutamate receptor (the AMPA receptor, see below). ACPD (trans-(±)-1-amino-1,3-cyclopentanedicarboxylic acid) appears to be a more specific synthetic agonist for the mGluR. Unfortunately, the lack of a specific antagonist of the mGluR has hampered the research of its function, especially with regard to its role in anoxic brain damage.

The ionotropic receptors are directly coupled to non-selective cation channels, and their activation leads to an inflow of Na^+ or Ca^{2+} that depolarizes the neuronal membrane.[69,95] There are three main types of ionotropic glutamate receptors, and each of there has been named after an agonist that more or less specifically activates it, viz. AMPA (amino-3-hydroxy-5-methylisoxazole-4-proprionic acid), KA (kainic acid) and NMDA (N-methyl-D-aspartate).

The AMPA receptor opens a channel with conductance for Na^+ and it appears to mediate most of the rapid synaptic excitatory neurotransmission in the brain. It could be mentioned that this receptor was originally called the quisqualate receptor, but this name was dropped after it was found that quisqualate also activates a metabotropic glutamate receptor, hence being less specific than AMPA.

Cloning experiments have demonstrated that KA and AMPA receptors are without doubt distinctly different molecular complexes.[5] However, the KA receptor is not as well studied as the AMPA or the NMDA receptor and its role in the brain is uncertain. It is activated by low levels of KA, but higher levels of kainic acid will also activate AMPA receptors, which has made it difficult to experimentally identify KA receptor mechanisms in brain. Often, the AMPA and KA receptors are lumped together under the term non-NMDA receptors. This is partly because of the difficulties in separating the roles of these receptors, and partly because of the similarities in action, both KA and AMPA receptors are ionotropic receptors that open up a channel with a high Na^+ conductance in response to their endogenous agonist, glutamate. Although non-NMDA receptors have generally been considered to be impermeable to divalent cations, evidence suggest that at least some sub-populations of AMPA and KA receptors are permeable for Ca^{2+}.[5]

As mentioned, glutamate is known to play a particularly detrimental role in the anoxic mammalian brain.[90,93] Much, if not most, of the damage caused by glutamate during anoxia appears to be mediated by an over-activation of the ionotropic NMDA receptor which opens a channel with conductance not only for Na^+ but also for Ca^{2+} (Fig 2.5). It is apparently the latter conductance that makes it so potentially harmful (the central role of Ca^{2+} in anoxic brain damage will be dealt with in Chapter 4).

The NMDA receptor appears to be the most widely distributed ionotropic glutamate receptor in the mammalian CNS[68]. It is also the most studied receptor, both because of its involvement in glutamate excitotoxicity[90], and because of its apparent role in long term potentiation of synaptic potentials, a mechanism believed to be associated with learning and memory[12].

Like the other ionotropic glutamate receptors, the binding site for glutamate on the NMDA receptor and the non-specific cation channel forms a transmembrane protein complex, hence, it is sometimes referred to as the NMDA channel. At the resting potential the NMDA receptor is normally blocked by Mg^{2+} (Fig. 2.5)[63]. However this block is removed when the membrane is depolarized. The NMDA receptor also contains a binding site for glycine[49] and another for polyamines like spermine and spermidine[86]. The binding of glycine and polyamines to the NMDA receptor facilitates the opening of the ion channel, and, in fact, glutamate is virtually without effect on the receptor unless the glycine modulatory site is also occupied. This glycine binding site is not blocked by strychnine, which shows that it is distinctly different from the glycine receptor occurring at high densities in the spinal cord and brain stem (discussed below). Ketamine, dizocilpine maleate (MK-801) and phencyclidine (PCP or "angels dust") are non-competitive antagonists of the NMDA receptor, binding to a site within the ion-channel and not on the sites where glutamate, glycine or polyamines bind. However, other exogenous antagonists act competitively by blocking the glutamate binding site[111], and there are also synthetic antagonist acting on the glycine and polyamine binding sites. Finally, to add to the complex pharmacology and endogenous modulation of this receptor, it could be mentioned that Zn^{2+} appears to be an inhibitory modulator of the NMDA receptor. As has repeatedly proven to be the case with neurotransmitter receptors, detailed ligand binding studies as well as molecular cloning now suggest that there are subtypes of the NMDA receptor.[101]

1.3.2 GABA

GABA is the major inhibitory neurotransmitter in the vertebrate brain. As much as 20–40 % of all vertebrate CNS neurons are estimated to be GABAergic[29,64] and virtually all neurons in the vertebrate CNS are sensitive to the application of GABA[52]. It has been found that glial cells possess GABA gated ion channels though their functional significance is presently unknown[52]. GABA also has peripheral actions in vertebrates and acts as an inhibitory transmitter in some non-neural cells within the endocrine system.[88]

Two major subtypes of GABA receptors in vertebrates were classified in the early 1980s – the $GABA_A$ and the $GABA_B$ receptor (Fig 2.6). The $GABA_A$ receptor is directly coupled to a monovalent anion channel and when GABA binds to the $GABA_A$ receptor, the Cl^- conductances of the neuronal membrane increases.[9] It has been shown that activated $GABA_A$ receptors also become permeable to HCO_3^- ions.[52] The binding site for GABA and the ion channel forms a single protein complex made up of 5 more or less different, subunits (i.e. a hetero-oligomer).[9] Some of these subunits have distinct binding sites for the substances known to modulate the $GABA_A$ receptor: barbiturates, benzodiazepines and steroids. Also certain cations exert a direct modulatory action on $GABA_A$ receptors. Intracellular Ca^{2+}, which can

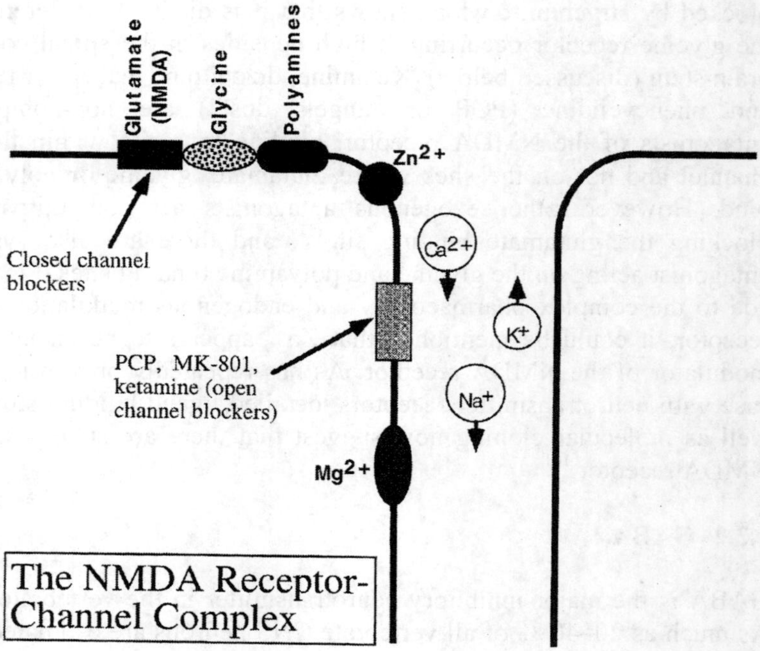

Fig. 2.5 The NMDA receptor-channel complex. Binding of glutamate (or NMDA) to the receptor opens a channel with conductance for Na^+ and Ca^{2+}. The binding site for glutamate and the cation channel forms a transmembrane protein complex. At the resting potential the NMDA receptor is normally blocked by Mg^{2+}, a block that is removed when the membrane is depolarized. The receptor also contains a binding sites for glycine and polyamines like spermine and spermidine. The binding of glycine and polyamines facilitates the opening of the ion channel. Ketamine, dizocilpine maleate (MK-801) and phencyclidine (PCP) are non-competitive antagonists. Zn^{2+} appear to be an inhibitory modulator of the NMDA receptor (redrawn from Olney[81]).

increase greatly during anoxia/ischemia, has significant modulatory actions on $GABA_A$ receptors. For example, rat hippocampal neurons show a clear Ca^{2+} dependent down-regulation of GABA-gated conductance.[11] $GABA_A$ receptors are also sensitive to extracellular pH, a decrease in pH_e generally causing an increase in GABA-gated Cl^- and HCO_3^- conductance[52]. A decrease in extracellular pH commonly occurs during anoxia/ischemia. Activation of the $GABA_B$ receptor causes the opening of a K^+ channel (Fig 2.6) and a decrease in Ca^{2+} conductance. $GABA_B$ receptor is G protein linked to the ion channels and some of its effects are probably also mediated by intracellular messenger mechanisms[8,27].Relatively recently, a $GABA_C$ receptor has been identified[50]. Like the $GABA_A$ receptor, it is a receptor gated Cl^- channel. However, it is a homo-oligomer, made up of identical subunits showing 30–38% homology with the subunits of the $GABA_A$ receptor. The $GABA_C$ receptor is insensitive to bicuculline (a well known $GABA_A$ receptor antagonist) as well as barbiturates, benzodiazepines and steroids, and the subunit of the $GABA_C$ receptor lacks binding sites for these substances. However, the $GABA_C$ receptor has an affinity for GABA 10 times higher than that of the $GABA_A$ receptor and it is less liable to desensitization upon stimulation with GABA. The $GABA_C$ receptor displays several pharmacological similarities to the $GABA_A$ receptor. Thus, muscimol, a potent $GABA_A$ agonist, is a partial agonist of the $GABA_C$ receptor. However, cis-4-aminocrotonic acid (CACA) appears to be a relatively specific agonist for the $GABA_C$ receptor. The best evidence for the occurrence of functional $GABA_C$ receptors comes from studies of the retina (of both mammals and fish). However, mRNA for subunits of this receptor has been found to be produced in all mammalian brain regions, suggesting a widespread distribution of $GABA_C$ receptors in the CNS.[50]

The binding of GABA to the GABA receptors will increase Cl^-, HCO_3^-, and K^+ conductance as well as decrease Ca^{2+} conductance and result in the membrane potential being clamped at the resting potential or even hyperpolarized. Since this will increase the excitatory input needed to open voltage-gated Na^+ and Ca^{2+} channels a release of GABA will diminish the rate of firing of energetically costly action potentials and will inhibit further neuronal excitation. However, if prolonged the increased ionic conductances may gradually lead to a dissipation of Cl^-, HCO_3^- and K^+ gradients and a shift towards depolarization. Indeed, in intensely activated dendritic $GABA_A$ receptors the depolarization caused by the loss of Cl^- and HCO_3^- gradients is sufficient to diminish the voltage-dependent magnesium block of the NMDA receptor with the result that neurons are excited rather than inhibited.[100]

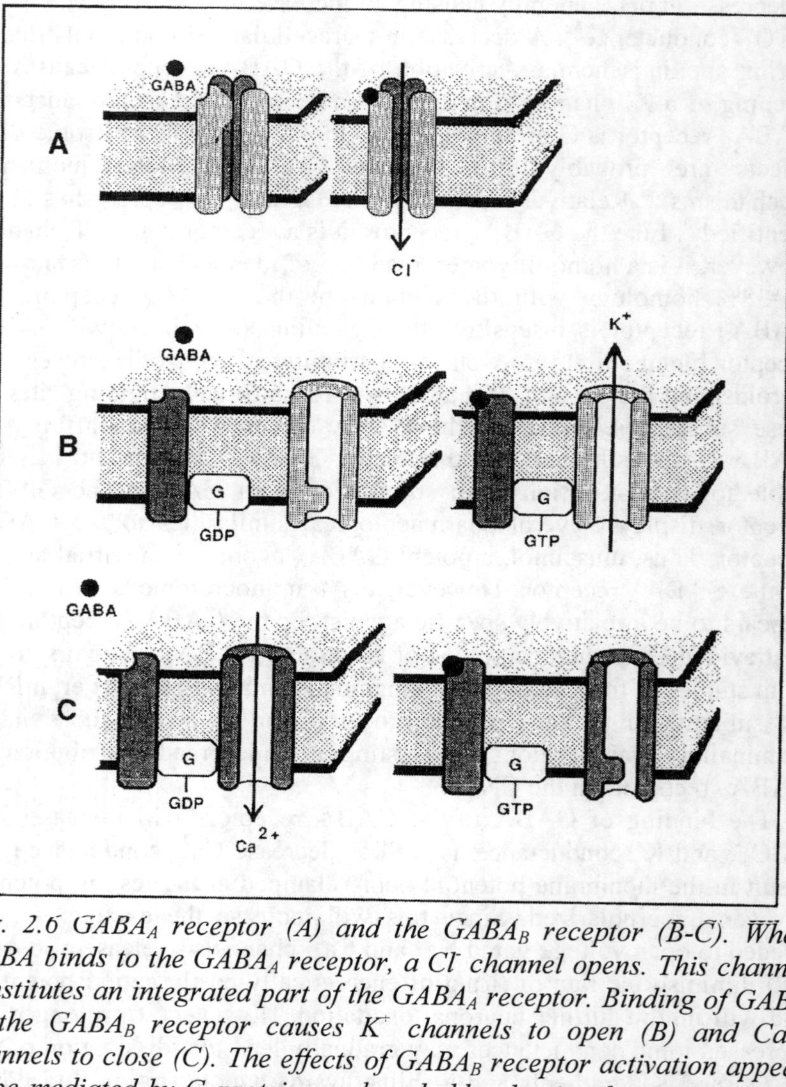

Fig. 2.6 GABA$_A$ receptor (A) and the GABA$_B$ receptor (B-C). When GABA binds to the GABA$_A$ receptor, a Cl^- channel opens. This channel constitutes an integrated part of the GABA$_A$ receptor. Binding of GABA to the GABA$_B$ receptor causes K^+ channels to open (B) and Ca^{2+} channels to close (C). The effects of GABA$_B$ receptor activation appear to be mediated by G proteins and probably also involves intracellular second messenger mechanisms. From Nilsson and Lutz.[77]

GABA is synthesized from glutamate (Fig. 2.7) via a single enzymatic step that is not dependent upon oxygen.[92] GABA break down, on the other hand, which is catalyzed by the enzymes GABA aminotransferase and succinic semialdehyde dehydrogenase, requires oxygen because it demands two substances that are in short supply during anoxia: NAD and the tricarboxylic acid-cycle intermediate ketoglutarate. Further, the synthesis of neuronal glutamate is mainly aerobic since, to a large extent, neuronal glutamate appears to be derived from ketoglutarate (partly via glutamine). The result is that both the resynthesis of glutamate and the break down of GABA will slow down or stop when the oxygen level falls. During anoxia or hypoxia therefore, these pathways will naturally tend to cause an increase in GABA and a decrease in glutamate. It can be imagined that their possible regulatory influences in reducing metabolism during times of oxygen deprivation are very ancient in the evolutionary scheme.[74]

1.3.3 Glycine

Glycine is well established as an inhibitory neurotransmitter in lower brain areas and the spinal cord, where activation of its receptor opens Cl^- channels. However, as mentioned, in the mammalian neocortex, glycine functions as an essential allosteric activator of glutamate receptors (NMDA receptors) thereby being involved in excitatory neurotransmission.[25,49]

1.3.4 Monoamines

The monoamines of the vertebrate brain include three extensively studied neurotransmitters – serotonin (5-hydroxytryptamine, 5-HT), dopamine and norepinephrine. Although vertebrate monoamine neurotransmitters appear to be involved in numerous neuronal processes, including locomotor control, central control of hormone release, central blood pressure control, and the modulation of social behavior[62,96,110], the number of monoaminergic neurons appears to be comparatively small. In fact, the mammalian brain only contains a few thousand monoaminergic neurons.

There is a great variety of monoamine receptors that by effecting different second messenger systems (cAMP, intracellular Ca^{2+}, and IP_3) can activate different ion channels. The result is that monoamines can have either excitatory or inhibitory effects. However, it seems that the predominating effects of 5-HT and norepinephrine are inhibitory, while dopamine mainly acts as an excitatory neurotransmitter on central neurons. For example, activation of the 5-HT_{1A} receptor causes a cAMP mediated opening of K^+ channels, the 5-HT_{1C} receptor opens a Cl^- channels (via IP_3). On the other hand, the 5-HT_3 receptors appear to belong to a ligand-gated receptor family where the binding site for the neurotransmitter is on the same protein complex as a nonspecific channel for monovalent cations (similar to the NMDA receptor) [30].Monoamines are particularly interesting in the context of anoxia tolerance because their synthesis and some steps in their metabolic degradation have an absolute demand for molecular oxygen (Fig. 2.8).

With regard to 5-HT, both its synthesis and degradation are oxygen dependent. In the first step of the 5-HT synthesis, L-tryptophan is hydroxylated by tryptophan hydroxylase. This reaction involves a simultaneous reduction of molecular oxygen. Monoamine oxidase (MAO) catalyses the first step of the degradation of 5-HT, in which an aldehyde intermediate is formed. This reaction also involves a simultaneous reduction of molecular oxygen. The aldehyde intermediate is then rapidly oxidized by NAD dependent aldehyde dehydrogenase to form 5-hydroxyindole-3-acetic acid (5-HIAA), or, to a minor extent, reduced by NADPH dependent aldehyde reductase to form an alcohol (5-hydroxytryptophol). Catecholamine synthesis resembles 5-HT synthesis, and has the same absolute requirement for molecular oxygen. In the first step of the synthesis of all catecholamines, L-tyrosine is hydroxylated by tyrosine hydroxylase to form L-dopa. The oxygen atom that is added to tyrosine in this reaction is directly derived from molecular oxygen.[4,57,72] Also the synthesis of norepinephrine from dopamine, which is catalyzed by dopamine-ß-hydroxylase, has a direct requirement for molecular oxygen.[57,72] With regard to catabolism, the monoamine oxidase catalyzed oxidative deamination of catecholamines demands molecular oxygen.[2,113] However, in contrast to the absolute oxygen demand by 5-HT catabolism, catecholamine inactivation could probably proceed in the absence of oxygen, via catechol-O-methyl-transferase (COMT) catalyzed methylation or by sulphate or glucuronide conjugation.[72] Hence, there are reasons to expect decreases in catecholamine levels during prolonged anoxia.

1.3.5 Acetylcholine

Early in the last century, its was discovered that there are two main types of receptors for the neurotransmitter acetylcholine, the nicotinic and the muscarinic receptor, both named after alkaloids (nicotine and muscarine) known to specifically block each of these receptors. With regard to the nicotinic receptor, the binding site for the neurotransmitter is on the same protein complex as a nonspecific channel for monovalent cations. Indeed, the nicotinic receptor shows many amino acid sequence homologies with the 5-HT_3 receptor, the $GABA_A$ receptor and the glycine receptor, and together, these form a family of ligand-gated ion channels. (The NMDA receptor for glutamate shows no overall homology with these receptors, but forms another receptor family together with other glutamate receptors.)

Activation of the muscarinic receptor can have a variety of cellular effects since it interact with G proteins that can inhibit adenylate cyclase, activate phospholipase C or open K^+ channels. Thus, the intracellular mediators that changes in concentration after muscarinic receptor activation can include cAMP, diacylglycerol and IP_3, which in turn may have effects such as release of Ca^{2+} from intracellular stores.

After having been released, the action of acetylcholine is terminated extracellularly, in the synaptic cleft, by the enzyme acetylcholinesterase. This is in contrast to most other neurotransmitters, which are inactivated by reuptake into neurons or other cells at the synapse.

1.3.6 Adenosine

Adenosine is produced during the enzymatic breakdown of the high energy purines ATP, ADP and AMP (Fig 2.9). It is formed from AMP, either directly (by 5'-nucleotidase) or indirectly via IMP and inosine. The source of AMP may be intracellular ATP as well as extracellular ATP, the latter released as a neurotransmitter (or co-released with another neurotransmitter, since ATP is sometimes present in the synaptic vesicles of other neurotransmitters).

Intracellularly, the most important enzymatic step acting to decrease the adenosine level appears to be the ATP dependent adenosine kinase reaction, which forms AMP from adenosine.During energy deficiency, when ATP consumption exceeds ATP production, there is a net increase in the AMP available for adenosine formation. At the same time, the ATP level falls, which will slow down the adenosine kinase reaction. Consequently, in the energetically compromised brain, adenosine levels increase substantially, both intracellularly and extracellularly.[2,73] In the mammalian brain, excessive neuronal activity as well as anoxia, ischemia and hypoglycemia results in substantially increased levels of adenosine.[23]

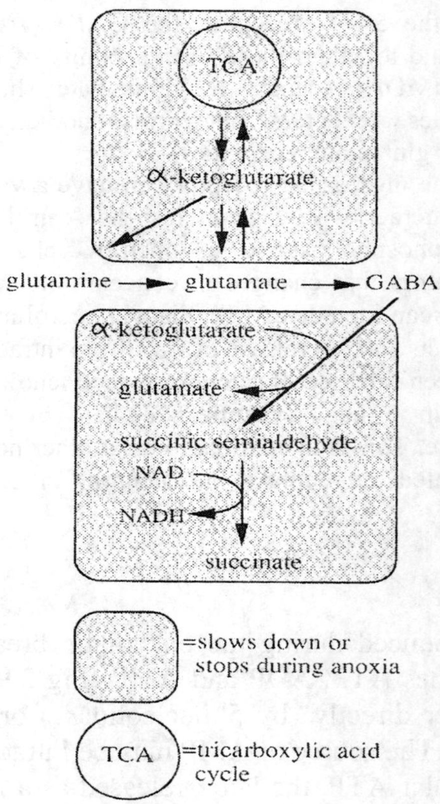

Fig. 2.7 The interrelated metabolism of neuronal GABA and glutamate. Note that GABA is synthesized from glutamate in a single enzymatic step that is not dependent upon oxygen. By contrast, the synthesis of glutamate and the breakdown of GABA are both inhibited by a fall in oxygen availability.

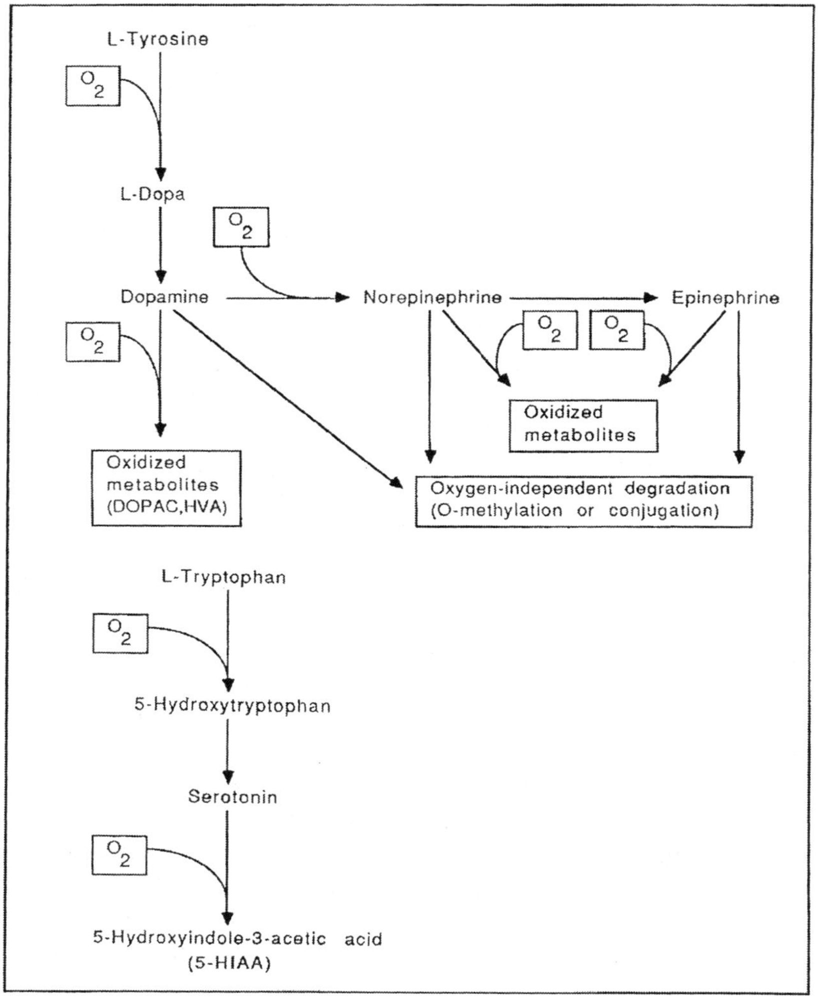

Fig. 2.8 Overview of monoamine metabolism. These neurotransmitters are of particular interest in the context of anoxia because their synthesis and some steps in their metabolic degradation have an absolute demand for molecular O_2.

Although extracellular adenosine has powerful effects on neurotransmission, adenosine does not appear to be stored in vesicles, like the classical neurotransmitters, and is therefore, generally referred to as a neuromodulator. Since adenosine lacks the charged phosphate groups, it is more hydrophobic than ATP, ADP or AMP, so it may diffuse out of the cells through the lipid cell membrane. However, most of the adenosine release appear to involve nucleotide transporters, which are directional (driven by the Na+ gradient) as well as equilibrative (driven by the adenosine gradient) (see Rudolphi et al.[91] Fredholm[23] for reviews).

Most of our knowledge of adenosine as a neuromodulator comes from mammalian studies, but adenosine receptors are much more widespread. They occur in reptilian and fish brain, and have been shown to inhibit neurotransmitter release in fish [82]. Because of its rapid release during incipient energy failure and its role as an inhibitory neuromodulator, adenosine is a very promising candidate as primary initiator of metabolic depression and as a promoter of anoxic survival. In the mammalian brain, activation of adenosine receptors causes a decrease in neuronal excitability (postsynaptic inhibition) and a suppression of neurotransmitter release (presynaptic inhibition).[92,100] thereby producing a decrease in energy utilization. In addition, adenosine release facilitates energy substrate supply to the mammalian brain by increasing cerebral blood flow[4,68] and by stimulating glycogenolysis.[61]

The adenosine receptors were long divided into two major types, the A1 and A2 receptors[99], but more recent evidence have clearly established the presence of an A3 receptor as well as two sub-types of the A2 receptor (A2A and A2B). All four of these adenosine receptors have been cloned and they are also distinguished on the basis of their differential interaction with a variety of adenosine analogues.[23] In addition, an A4 receptor has been suggested to exist.

The A1 and A2 receptor types are blocked by methylxanthines such as caffeine, theophylline or aminophylline (a dimer of theophylline that is rapidly cleaved in tissues). Several derivatives of these xanthines have proven to be more specific antagonists for the different receptor sub-types, while derivatives of adenosine are used as sub-type specific agonists.

The effects of adenosine on reducing neuronal excitability and neurotransmission appear to be mediated by the A1 adenosine receptors,

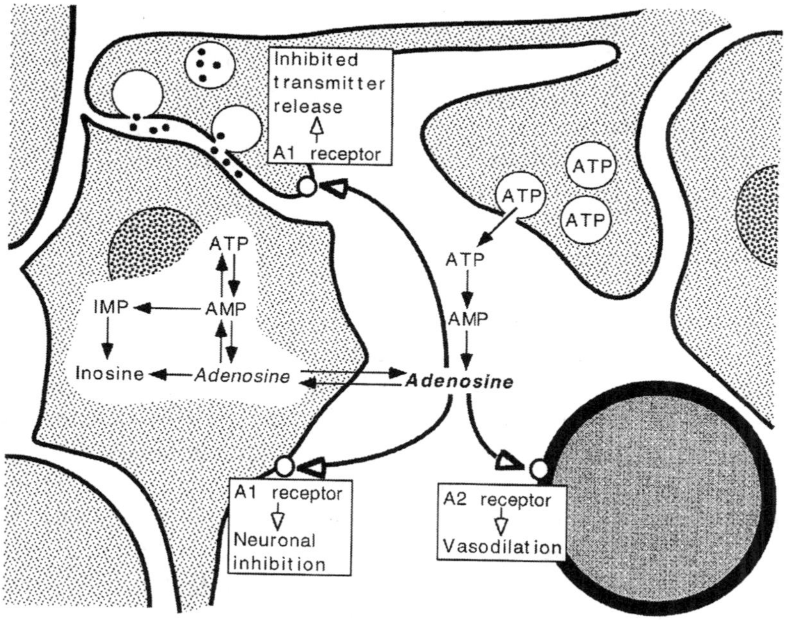

Fig. 2.9 Adenosine in brain: its effects and its metabolism. Adenosine can be produced from intracellular ATP break down caused by anoxic/ischemic energy deficiency or neuronal over-excitation. It can also be derived from ATP released as a neurotransmitter (or co-transmitter). When present extracellularly, adenosine activates A1 receptors, causing a decrease in neurotransmitter release and neuronal excitability. In addition, it binds to A2 receptors, causing vasodilation and, thereby, a local increase in cerebral blood flow.

while the vasodilatory effects of adenosine operates via the A2 receptors. There is also evidence, however, that stimulation of A2 receptors can have excitatory effects.[103] In most tissues, activation of A1 receptors reduce adenyalte cyclase activity, and, thus, cAMP production, while A2 receptors stimulate adenylate cyclase. In both cases, the receptors appear to be coupled to adenylate cyclase via a G protein. A1 receptor activation has also been found to open K^+ channels, and possibly also regulate Ca^{2+} channels. Clearly, the opening of K^+

channels must underlie some of the inhibitory effects of adenosine on nervous activity.

Like mammals, adenosine has been shown to inhibit neurotransmitter release in fish brain[82], acting through A1 receptors.[89] The latter study also indicated the presence of A2B receptors in fish brain (goldfish). Adenosine superfused onto the brain of crucian carp[76] or freshwater turtles[41], causes a substantial increase in brain blood flow occurs, indicating cerebral vasodilation.

1.3.7 Nitric oxide (NO)

In 1992, NO, a small gaseous molecule and free radical, was declared by the journal Science to be "the molecule of the year". The reason was simple: during the five years since 1987, when Palmer and coworkers[83] showed that NO was a biological messenger involved in vasodilation, this unlikely molecule had been found to participate in almost all types of physiological process, including several aspects of neuronal communication in brain, macrophage function, platelet aggregation, penile erection and vasodilation.[105,114] In 1998, a Nobel prize was awarded jointly to Robert F. Furchgott, Louis J. Ignarro and Ferid Murad for their discoveries concerning NO as a signalling molecule in the cardiovascular system. It has become increasingly clear that NO also plays an important role in the pathophysiological events[8], including those that take place in the anoxic or ischemic brain.[16]

Because NO is a gas that readily penetrates biological membranes, and because it is highly reactive with a half life of a few seconds, it cannot be stored in vesicles. Therefore its concentration and release is regulated at the level of biosynthesis. NO is synthesized by nitric oxide synthase (NOS), an enzyme which occurs in several isoforms. However, regardless of isozyme form, the reaction catalyzed is the same.

$$\text{L-arginine} + O_2 \xrightarrow{\textbf{NOS}} \text{NO} + \text{L-citrulline}$$

L-arginine is the universal substrate for NOS and the reaction produces not only NO but also L-citrulline. NADPH, FAD (flavine adenine dinucleotide), FMN (flavine mononucleotide) and BH_4 (tetrahydrobiopterine) are cofactors in this reaction. In the context of this book, it is of special interest to note that molecular oxygen is consumed in this process.[58] However, although NO cannot be produced in the complete absence of oxygen, NO production is very much involved in the effects that hypoxia (low oxygen) and ischemia have on the brain, as we shall see in Chapter 4.

There are two main groups of NOS isozymes: constitutive NOS and inducible NOS. The NOS occurring in neurons (neuronal NOS or NOS-I) and that of endothelial cells in arteries (endothelial NOS or NOS-III) belongs to the constitutive group. Their activity is rapidly turned on by a increase in intracellular Ca^{2+}, a mechanism mediated by the binding of Ca^{2+} to calmodulin, which in turn binds to a calmodulin binding site on NOS. Constitutive NOS isoforms can only generate relatively small amounts of NO (picomoles) for short periods of time. It should be mentioned that the neuronal and endothelial NOS isoforms are not as specifically distributed as their names imply. For example, NOS-III occur in CA1 neurons of the mouse hippocampus.[80] The inducible form of NOS (NOS-II) is, for example, found in macrophages, where it mediates their antimicrobal action. NOS-II also appears to exist in endothelial cells[43] and glia cells[22]. Inducible NOS is insensitive to Ca^{2+}. It is known to be activated by cytokines and endotoxins, a process that requires DNA transcription and synthesis of the enzyme. Thus, it can only slowly be turned on (hours or days), but in contrast to constitutive NOS, it can release large amounts of NO for long periods (days).

After being produced, NO diffuses freely into surrounding cells. In neurons as well as arteries, a predominant first effect of NO is the activation of guanylate cyclase which catalyses the formation of cGMP. Subsequent physiological actions are then initiated by the increased levels of cGMP. In neurons and endothelial cells, the increase in intracellular Ca^{2+} that stimulates NOS activity can be initiated by activation of various membrane receptors. Thus, the binding of glutamate to its NMDA receptor can stimulate NO production in neurons (Fig. 2.10). In endothelial cells, NOS is known to be activated by receptors for a rather long array of vasoactive substances, including acetylcholine, ATP, histamine, 5-HT, noradrenaline, endothelin, substance P, arginine vasopressine, bradykinin, oxytocin, and prostaglandin F_{2q} (Fig 2.10).[22] With regard to endothelial cells, it has been suggested that mechanical (shear) forces associated with the passing blood can also stimulate NO synthesis.[28] NO synthesized in endothelial cells diffuses into the surrounding arterial smooth muscle where it activates a soluble form of guanylate cyclase that catalyses cGMP production. The increased levels of cGMP cause relaxation of the smooth muscle and, hence, vasodilation and increased blood flow. Since NO diffuses freely in tissues and may reach targets several hundred micrometers from the point of synthesis, several sources of NOS, other than endothelial NOS have the capacity to cause cerebral vasodilation (Fig. 2.10). Many arteries and arterioles in the brain are innervated by NOS containing neurons. NO released from nerve-endings of so called non-adrenergic non-cholinergic neurons (NANC neurons) has been suggested to participate in the regulation of cerebral blood flow.[104] Other possible sources of NO that may mediate cerebral vasodilation are glial cells, macrophages, and platelets. However, for vasodilation, the importance of other NO sources than the endothelium might be relatively minor. Experiments on mice, with the gene for neuronal NOS knocked out, suggest that at least the vasodilatory effect of topical acetylcholine administration is caused by endothelial NOS rather than neuronal NOS.[44] Acetylcholine has been found to cause NO dependent vasodilation not only in the brain of mammals but also in that of anoxia tolerant vertebrates such as the crucian carp[40] and freshwater turtle[42] – two animals that will be very much in focus later on in this book.

The development of potent NOS inhibitors such as such as N^G-monomethyl L-arginine (L-NMMA), N^G-nitro-L-arginine (L-NA) and N^G-nitro-L-arginine methyl ester (L-NAME)[43,45,67] has been of tremendous importance for the experimental search for NOS dependent mechanisms. A blocking effect of a NOS inhibitor on the physiological phenomenon studied provides evidence for an involvement of NO. Drugs that can selectively block specific isoforms of NOS are now being developed. L-MIN and ARL-17477 appear to be specific blockers of neuronal NOS[6], while aminoguanine is probably relatively specific in it inhibitory action on inducible NOS.[26,97] Such specific inhibitors may be particularly useful in the context of hypoxia/ischemia.

Sodium nitroprusside (SNP), S-nitroso-N-acetyl-penicillamine (SNAP), and 3-morpholinosydnonimine (SIN-1) are other pharmacological tools that are regularly used to find indications of NO dependent mechanisms. These substances decomposes in tissues to form NO, reactions that does not require NOS.NOS activity has been detected in virtually every major part of the brain, not only in mammals but also in birds, reptiles, amphibians and fish.[36,108] A fortunate circumstance that have greatly aided histochemical detection of neuronal NOS is that this enzyme reduces nitro-blue tetrazolium to produce a clearly visible stain (a reaction that NOS catalyzes also in formaldehyde fixed tissue). This reaction is called the NADPH diaphorase reaction and in fixed tissue it appears to be highly specific for neurons containing NOS, since in transgenic mice lacking a functional gene for NOS, no NADPH diaphorase activity can be detected in brain.[37] In addition to the NADPH diaphorase reaction method, also antibodies raised against the various forms of NOS have been used to map the distribution of NOS in brain.

1.3.8 Adenosine-Tri-Phosphate (ATP) as a Signalling Molecule

ATP release has been widely observed. It appears to be the principle neurotransmitter in purinergic cells as well as being copackaged with serotonin and noradrenaline in other neuronal systems.[109] ATP can also cross the neuronal membrane in a nonexocytotic manner and is released from damaged cells following such insults as ischemia. There is considerable evidence that ATP can act as an excitatory neurotransmitter in the mammalian brain.[17,21] ATP induced excitation is mediated by the binding of ATP to P_2 purinoreceptors, a class of receptors that can be blocked by the drug suramin. The ATP P_2 receptors have been subdivided, and one in particular, P_{2x}, appears to be responsible for ATP activated fast excitatory synaptic transmission by acting as a non-selective cation channel with high permeability for Ca^{2+}, Na^+ and K^+.[17] Unlike the NMDA receptor, Ca^{2+} influx through P_2 receptors does not appear to be voltage dependent and may therefore occur in absence of cell depolarization.[17]

Extracellular ATP may also function as an endogenous regulator of regional cerebral blood flow, acting to dilate cerebral blood vessels.[24] The effects of released ATP are likely to be transient as extracellular ATP is rapidly hydrolyzed by ecto ATPases to adenosine. Since adenosine is an inhibitory neuromodulator the following interesting scenario is possible: initially ATP release results in rapid excitation, but the effect is only transitory and is followed by a delayed and longer lasting inhibition through the action of adenosine. Like glutamate, an excessive release of ATP might have potential neurotoxic effects. But while hypoxia/ischemia produces an increase in ATP release in cardiac tissues[7], Phillis et al.[84] were not able to detect an increase in cortical ATP in the cerebral cortex of hypoxic rats (8% oxygen inhalation).

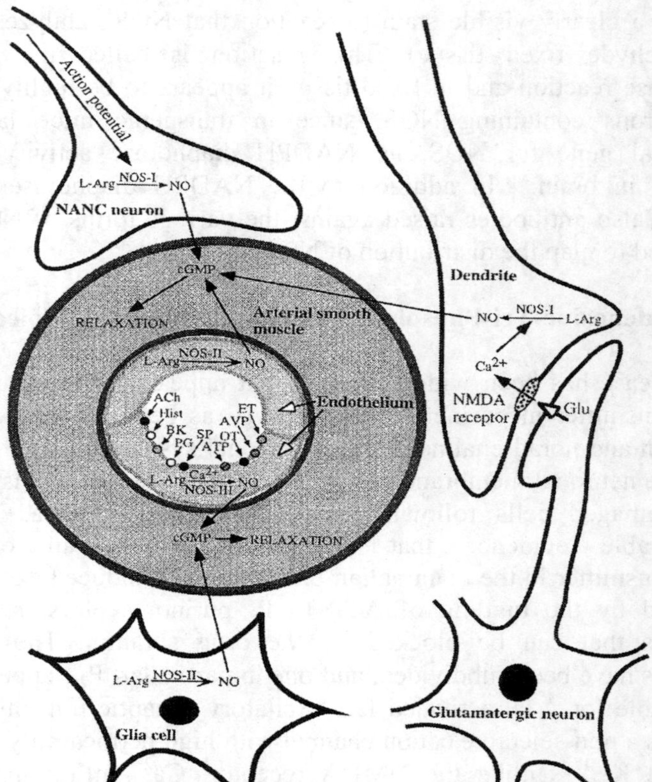

Fig. 2.10 Mechanisms behind NO mediated arterial vasodilation in the mammalian brain. Note that there are several possible sources of NO: the endothelium, glia cells and different types of neurons. Among the neurons, the non-adrenergic non-cholinergic (NANC) neurons appear to be mainly directed at stimulating vasodilation. The endothelial NOS can be activated by several vasoactive substances present in the blood, including acetylcholine (ACh), Histamine (Hist), bradykinin (BK), prostaglandin F_{2q} (PG), substance P (SP), ATP, oxytocin (OT), arginine vasopressin (AVP) and endothelin (ET). Like glia cells, endothelial cells also contain inducible NOS (NOS-II). When NO reaches the arterial smooth muscle cells, it activates guanylate cyclase that catalyses the production of cGMP, which causes relaxation of the muscle. From Nilsson and Söderström.[78]

1.3.9 ATP-sensitive K^+-channels

ATP-sensitive K^+ channels (K_{ATP} channels) open in response to low cellular ATP concentrations. As discussed above, the opening of K^+ channels will tend to clamp the membrane potential and counteract neuronal excitation. Thus, these channels directly respond to the intracellular energy charge, and may be intimately involved in maintaining energy homeostasis. K_{ATP} channels were first described for mammalian cardiac muscle[79], and they were soon shown to play a pivotal role in insulin secretion by pancreatic ß-cells.[13] The existence of K_{ATP} channels was established in the mammalian central nervous system.[1] There is very little data on K_{ATP} channels from non-mammalian vertebrates. Jonas et al.[51] found evidence for a K_{ATP} channel in peripheral axons of the clawed frog (*Xenopus laevis*) that could be inhibited by glibenclamide, a potent blocker of mammalian K_{ATP} channels. However, Jiang et al.[48] obtained results suggesting that there are no (or very little) glibenclamide blockable K_{ATP} channels in the brain of freshwater turtles.

In the mammalian brain and heart, K_{ATP} channels are thought to be important participants in the regulation and maintenance of the cellular energy status. During periods of low activity, when ATP demand is low and ATP concentrations high these channels will be closed and as long as ATP supply matches ATP demand, they will remain closed as the work load increases. However, as soon as the ATP demand outstrips the supply the ATP concentration will fall and cause an opening of the K_{ATP} channels. As discussed above, the resultant increase in membrane K^+ conductance clamps the membrane potential and puts a brake on action potential producing depolarization. In capillary smooth-muscle cells the same effect produces vasodilatation which results in an increased blood supply to tissues that have been energetically compromised.[15]

2. BRAIN ENERGY CONSUMPTION

In comparison to most other tissues, vertebrate brains have much higher rates of obligatory energy consumption and are correspondingly more susceptible to energy failure. Indeed it is well established that energy deficiency is *the* primary factor that initiates the chain of catastrophic events that occur in the anoxic brain.

Normally, glucose is the only fuel used by the brain for ATP production, although during periods of starvation it may also use ketone bodies. In contrast to most other tissues, amino acids or fatty acids make very minor contributions to brain energy production.[34]

The brain is highly aerobic. In the normoxic mammalian brain, more than 95% of the ATP is derived from aerobic glucose oxidation, and under fully aerobic conditions, lactate production accounts for less than 4% of the glucose metabolized.[20]

In this book, we use the term energy consumption as a synonym for ATP consumption. Energy consumption is a widely used expression in physiology, and all physiologists are of course aware of the fact that energy is not consumed, strictly speaking, but merely transformed into another form.

2.1 *Components of Energy Use*

The exceptionally high specific rate of energy consumption of the brain is related to the major role of the neurons in maintaining an electrically active state for the continual transmission of neuronal signals. Already in 1926, A. V. Hill found that stimulation of the electric activity in nerves correlated with increased oxygen consumption and heat production. Later, Yarowsky et al.[112] demonstrated a linear relationship between the rate of glucose utilization of the superior cervical ganglion and the frequency by which this ganglion is artificially stimulated. In fact, stimulating the superior cervical ganglion at a frequency of 15 Hz caused a doubling of the glucose consumption rate compared to the unstimulated control situation.

In this context the nerve cell can be seen as a leaky ionic battery, where the task of constantly recharging is performed by ATP demanding ion pumps. The most important pump related energy consumer is the Na^+/K^+ ATPase, which, apart from keeping up Na^+ and K^+ gradients, is also indirectly responsible for maintaining the Ca^{2+} gradient because part of the energy invested in the Na^+ gradient is used by a neuronal Ca^{2+}/Na^+ exchanger to exclude Ca^{2+} from the neurons. In addition, there are also Ca^{2+} pumps that directly consume ATP, although these are probably responsible for only a minor part of the brain energy consumption.[19]

Experimental evidence suggests that a major part, around 50-60%, of the total energy consumed by the brain is used for the extensive and continuous ion pumping needed to maintain ion gradients and, hence, to pay the metabolic costs of the endless opening of ion channels involved in the electric and synaptic activity.[20,33] This figure derives from studies on electrically depressed brains. If a mammalian brain is made electrically silent by deep anesthesia, its energy consumption falls to less than half of that of an awake brain[2,33,65], suggesting that more than half of the energy consumed by the brain is directly devoted to the electrical activity. Moreover, during coma, a condition characterized by a strongly suppressed electric activity, cerebral respiration is similarly reduced by about 50%.[94] Furthermore, the oxygen consumption of *in vitro* brain slices, which typically show only trivial spontaneous electric activity, is generally only 50 % of the corresponding *in situ* value.[59] Additional support comes from studies using ouabain, a specific blocked of the Na^+/K^+ pump. Applying ouabain to synaptosomes (isolates of nerve terminals) results in a very substantial reduction in their oxygen consumption (-43.5% turtle, -88.6% rat).[18]

Another set of evidence for the high energy demand of electric activity comes from metabolic measurements during epileptic seizures, where there is an uncontrolled elevation of electric activity. During seizures the rates of glucose utilization and lactate production may increase as much as 4 and 10 times, respectively, while the rate of oxygen consumption may rise 3-fold (reviewed by Erecinska and Silver[19]).

The remaining part of the energy consumed, i.e. that which is not directly related to electric activity, is presumably spent on such processes as counteracting basal ion leakage, neurotransmitter metabolism (including packing and transport), phosphorylation, and synthesis of structural molecules and enzymes. Unfortunately, very little is known about how much energy is consumed by each of these processes in brain or what happens to them when the brain's energy supply is compromised.

However, with regard to the energy consumed by leak currents, it is possible to make some comments. At the resting membrane potential the K^+ leak channels remain open with the result that the neurons are much more permeable to K^+ than to other ions (although the permeability to K^+ is also limited). However, since K^+ is close to its Nernst equilibrium potential the driving force for K^+ to move down its concentration gradient is small and the energy consumed by the K^+ leak current is very limited, at least compared to the energy consumed by the electric activity (see above). Indeed, by blocking both electric activity (by deep anesthesia) and the Na^+/K^+ pump (by ouabain), Astrup et al.[2] obtained results suggesting that the energy consumed by leak fluxes in the mammalian brain account for less than 25% of the total energy consumption of the awake mammalian brain. In the brain of an ectothermic vertebrate, the contribution of ion leakage to the total energy consumption is probably even less, because membranes of ectothermic vertebrates appear to display just a fraction of the leakiness of mammalian cell membranes.[39] One explanation for the higher leakiness of mammalian membranes is that the heat produced by leak fluxes is important in mammalian thermogenesis. No such function, of course, would be needed in "cold blooded" animals.

It is important in our discussions on brain energy failure to keep in mind that the energy consumption by the brain shows both striking regional and temporal differences. Six to nine-fold differences between the rate of glucose consumption in different brain areas have been found in rats[94] and monkeys[56], with the highest and lowest rates found in areas with gray and white matter, respectively. Spatio-temporal differences in energy consumption have also come in focus with positron-emission tomography (PET), where regional changes in glucose utilization can be continuously monitored *in vivo*. In humans, for example, motoric as well as mental tasks have been shown to correlate to regional changes in glucose metabolism in brain. Of course, the fact that these changes can be seen suggest that a considerable part of the energy consumed by the brain is devoted to neuronal communication.

2.2 *Rates of Energy Consumption in Vertebrate Brains*

Because of its relatively large size, constituting 2% of the body mass, the human brain is responsible for a very impressive portion of the body's energy consumption, viz., 20% of the total body oxygen consumption – a figure that has long been thought to make the human brain outstanding among vertebrates.

However, the size of the human brain is in no way unique (Fig. 2.11). In absolute size it is surpassed by that of dolphins, whales and elephants. Moreover, there are several small endothermic (warm-blooded) vertebrates, notably bats, mice, swallows, crows and sparrows, that have brains which make up 2.6-3.7% of the body weight.[14,66] Even greater values are seen in some primates, with the record holder being a small South American monkey (*Saimiri sciureus*) which has a brain that makes up 6.45% of the body weight.[107] The very high metabolic rates of these small endotherms make their brains comparatively less costly (using 4-13% of the body energy budget only), because the mass-specific scaling factor for brain energy consumption is much lower than that for whole body energy consumption in vertebrates.[66]

It was recently shown that the human brain is not outstanding even in having a brain that consumes the largest portion of the whole body energy budget. This record is currently held by the elephant nose fish (*Gnathonemus petersii*), a small (5-20 g) freshwater fish from tropical Africa with an unusually large brain: 3% of the body weight – a very impressive figure for a fish. Measurements of the rate of brain slice and body oxygen consumption in this species suggest that the brain accounts for some 60% of the whole body oxygen consumption during rest.[75] The elephant-nose fish, like the other members of the elephant-nose fish family (Mormyridae), are all so called weakly electric fishes, which means that they generate and perceive low current electric impulses.[67] This modality gives these nocturnal fishes the ability to orient in complete darkness, and they also use the system for communication with other individuals of the same species. Most, if not all, mormyrids seem to have large brains, and most of their brain is made up by a "gigantocerebellum" that controls their electric sense.

The elephant nose fish is a hypoxia tolerant species in the sense that it can tolerate low levels of oxygen in the ambient water – down to 10% of air saturation.[75] However, it does this by being exceptionally good at taking up oxygen from the water as well as by gulping air at the surface. These adaptations are likely to be essential for protecting the brain from hypoxic damage, since hypoxia is a common event in the Central African river systems it inhabit. In fact, if the water is made anoxic and the fish is kept away from gulping air, it rapidly loses its ability to co-ordinate its movement, suggesting that the nervous system has become energetically compromised. Thus, it is not in the possession of a hypoxia tolerant brain like that found in crucian carp or freshwater turtles – animals that will be in focus later in this book.

The energy consumption rate of the human brain on a per gram basis, the mass specific oxygen consumption, is also not in any way outstanding compared to other vertebrates. Contrary to the general belief, although whole body energy expenditures of ectothermic vertebrates are only about 10-20% of those of endotherms (at the same temperature and body size), when temperature is taken into account[60], the specific energy consumption of brain tissue of ectothermic vertebrates are not very different from that of the mammalian brain.[75] The average brain oxygen consumption rate of the three fishes shown in Fig. 2.12 (elephant nose fish, crucian carp and trout) is 4.95 mg g^{-1} h^{-1} (37 °C), which is not much lower than the 7.74 mg g^{-1} h^{-1} expected for an endothermic brain of the same average size (deducted from the regression line for endothermic brains in Fig. 2.12).[75] In fact, the oxygen consumption rate of the trout brain is actually on the regression line for endothermic brains. Moreover, Edwards et al.[18] found that at the

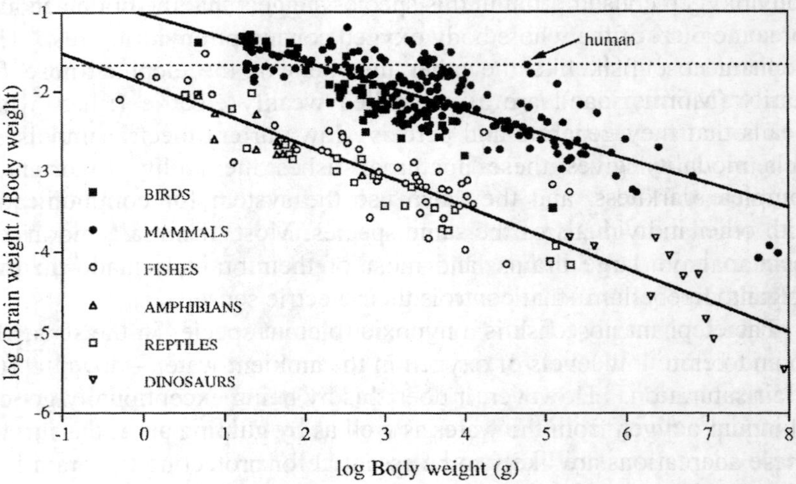

Fig. 2.11 The relative brain size in endothermic (warm-blooded) and ectothermic (cold-blooded) vertebrates. There is a fairly linear relationship between the logarithms of relative brain size (brain weight/body weight) and body weight. In general, the brain of and endothermic vertebrate is about 10 times bigger than that of an ectothermic vertebrate of the same size. The dotted line shows the relative brain size of the human brain (2% of the body weight). Note that there are numerous small mammals and birds, as well as one fish (the elephant nose fish), that have relative brain sizes that exceed that of man. Values are from Von Bonin[107], Crile and Quiring[14], Jerison[47] and Nilsson.[75]

same temperature (31 °C) rat and turtle brain synaptosomes consume identical amounts of oxygen and Robin et al.[87], could find no difference in the oxygen consumption rates of rat and turtle brain slices incubated at 21 °C. Caliguri et al.[10] reported that at an incubation temperature of 24 °C (pH 7.5) the oxygen consumption of the rat brain slice was only 64% greater than that of the turtle, compared to a 37 fold difference in metabolic rates between rat and turtle heart slices.

In addition, enzymatic studies suggest a similar metabolic intensity in the brains of ectothermic and endothermic vertebrates. In comparing homologous regions of the brain, Suarez et al.[102] found that the Na^+/K^+ ATPase activities were only 2-2.5 fold higher in rat than in turtle brain. Studies on lizard and rat mitochondria have suggested that these species have similar cytochrome oxidase activities. Thus, while the cytochrome oxidase activity of the lizard brain is about 72% that of the mammal when expressed on a tissue wet weight basis, on a per mg protein basis reptilian brain cytochrome oxidase activity is about 1.4 times greater than that of the mammal.[38] Also, when normalized for temperature, rat and turtle brains have equivalent maximal activities of citrate synthase (a measure of the capacity for oxidative metabolism), while the turtle has higher levels of lactate dehydrogenase and hexokinase, two glycolytic enzymes.[102]

The reason why ectothermic brains do not differ much from endothermic brains with regard to the rate of energy utilisation is probably related to the fact that ionic leakage, which is much higher and largely coupled to heat production in endothermic vertebrates[38,39], plays only a minor role in the overall energy expenditure of the brain. The ion movements involved in electric and synaptic activity are responsible for the largest portion of the energy used by the brain[19], and in this respect there is probably little difference between endothermic and ectothermic brains. It is likely that ion leakage makes up a larger portion of the energy consumption in most tissues other than brain, which explains why there is a large difference in metabolic rate between similarly sized ectotherms and endotherms on the whole body level, while such a difference is virtually absent in brain. The fact that the anoxic rainbow trout loses its brain ion homeostasis as fast as the anoxic mammalian brain[75] is most likely a reflection of this similarity of brain energy use.

Thus, with regard to the whole body energy budget it is comparatively more expensive for an ectothermic vertebrate, with its energetically inexpensive body, to have a large brain. This may be a reason why most ectothermic vertebrates have small brains – generally about 1/10th of the size of endothermic vertebrates of the same size (Fig. 2.11). The result of this is a strikingly constant ratio of brain to body oxygen consumption in both ecto- and endothermic vertebrates – their brain being responsible for 1.5 to 8.5% of the body oxygen consumption.[66] The notable exceptions to this rule are apes (with brains consuming up to 13% of the oxygen taken up by the body), man (20%), and of course the elephant nose fish (60%) discussed above. By contrast, elasmobranchs (sharks and rays), which have comparatively large brains, are probably not exceptions to this rule since their brains appear to have unusually low weight specific rates of energy consumption (see the discussion on the epaulette shark in Chapter 8).

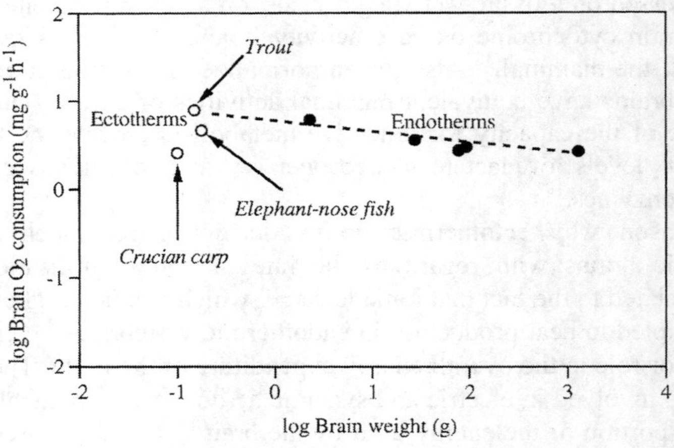

Fig. 2.12 Specific brain energy consumption (oxygen consumption) versus brain weight in endothermic and ectothermic vertebrates. Note that the specific brain energy consumption in ectotherms is similar to that in endotherms (rat, cat, dog, macaque and human). The values shown are those compiled or measured by Nilsson.[75]

In summary, the uniformity in brain metabolic demand across the vertebrates probably reflects similar costs for the continuous functioning of common basic processes – such as maintaining and restoring ionic gradients disturbed by the firing of action potentials, the synthesis and transport of neurotransmitters, and cell repair. These costs would not be expected to differ between endothermic and ectothermic vertebrates.

3. REFERENCES

1. Ashford ML, Sturgess NC, Trout NJ, Gardner NJ, Hales CN. Adenosine 5'-triphosphate-sensitive ion channels in neonatal rat cultured central neurons. Pflügers Arch 1988; 412:297-304.
2. Astrup J, Sorensen PM, Sorensen HR. Oxygen and glucose consumption related to Na^+-K^+ transport in canine brain. Stroke 1981; 12:726-730.
3. Baskys A. Metabotropic receptors and 'slow' excitatory actions of glutamate agonists in the hippocampus. Trends Neurosci 1992; 15:92-96.
4. Berne RM, Rubio R, Curnish RR. Release of adenosine from ischemic brain: Effect on cerebral vascular resistance and incorporation into cerebral adenine nucleotides. Circ. Res. 1974; 35:262-271.
5. Bettler B, Mulle C. Review: Neurotransmitter receptors II, AMPA and kainate receptors. Neuropharmacol. 1995; 34:123-139.
6. Bolanos JP, Almeida A. Roles of nitric oxide in brain hypoxia-ischemia. Biochim Biophys Acta 1999; 1411:415-436.
7. Borst M, Schrader J. Adenine nucleatide release from isolated perfused guinea pig hearts and extracellular formation of adenosine. Circ Res 1991; 68:797-806
8. Bowery N. $GABA_B$ receptors and their significance in mammalian pharmacology. Trends Pharmacol. Sci. 1989; 10:401-407.
9. Burt DR, Kamatchi GL. $GABA_A$ receptor subtypes: from pharmacology to molecular biology. FASEB J. 1991; 5:2916-2923.
10. Caliguri MA, Robin ED. Prolonged diving and recovery in the freshwater turtle IV. Effects of profound acidosis on O_2 consumption in turtle vs. rat (mammalian) brain and heart slices. Comp. Biochem. Physiol. A 1985; 81:603-605.
11. Chen QX, Stelzer A, Kay AR, Wong RK. $GABA_A$ receptor function is regulated by phosphorylation in acutely dissociated guinea-pig hippocampal neurones. J Physiol Lond 1990; 420:207-221.
12. Collingridge GL, Bliss TVP. Memories of NMDA receptors and LTP. Trends Neurosci. 1995; 18:54-56.
13. Cook DL, Hales CN. Intracellular ATP directly blocks K^+ channels in pancreatic ß-cells. Nature 1984; 311:271-273.
14. Crile G, Quiring DP. A record of the body weight and certain organ a nd gland weights of 3690 animals. Ohio J. Sci. 1940; 40:219-259.
15. Daut J, Maier-Rudolph W, von Beckerath N, Mehrke G, Gunther K, Goedel-Meinen L. Hypoxic dilation of coronary arteries is mediated by ATP-sensitive potassium channels. Science Washington DC 1990; 247:1341-1344.
16. Dawson DA. Nitric oxide and focal cerebral ischemia: multiplicity of actions and diverse outcome. Cerebrovasc. Brain Metab Rev 1994; 6:299-324.
17. Edwards FA, Gibbs AJ. ATP – a fast neurotransmitter. FEBS 1993; 325:86-89.
18. Edwards R, Lutz PL, Baden D. Relationship between energy expenditure and ion channel function in the rat and turtle brain. Am J Physiol 1989; 255:R1345-R1359.
19. Erecinska M, Silver IA. ATP and brain function. J Cereb Blood Flow Metab 1989; 9:2-19.
20. Ericinska M, Silver IA. Ions and energy in mammalian brain. Prog Neurobiol 1994; 43:37-71.

21. Evans R, Derkach V, Surprenant A. ATP mediates fast synaptic transmission in mammalian neurones. Nature 1992; 357:503-505.
22. Faraci FM, Brian JE. Nitric oxide and the cerebral circulation. Stroke 1994; 25:602-703.
23. Fredholm BB. Modulation of neurotransmitter release by heteroreceptors. In: Neurotransmitter Release and its Modulation. Powis DA, Bunn SJ (eds). Cambridge: Cambridge Univ. Press, 1995: 104-121.
24. Forrester T, Harper AM, MacKenzie ET, Thomson EM. Effect of adenosine triphosphate and some derivatives on cerebral blood flow and metabolism. J Physiol 1979; 296:343-355.
25. Foster AC, Kemp JA. Glycine maintains excitement. Nature 1989; 338:377-378.
26. Fukuto JM, Chaudhuri G. Inhibition of constitutive and inducible nitric oxide synthase: potential selective inhibition. Annu Rev Pharmacol Toxicol 1995; 35:165-194.
27. Gage PW. Activation and modulation of neuronal K^+ channels by GABA. Trends Neurosci 1992; 15:46-51.
28. Gaw AJ, Bevan JA. Flow-induced relaxation of the rabbit middle cerebral artery is composed of both endothelium-dependent and independent components. Stroke 1993; 24:105-110.
29. Gottlieb DI. GABAergic neurons. Sci Am 1988; February:38-45.
30. Göthert M. 5-Hydroxytryptamine receptors, an example for the complexity of chemical transmission of information in the brain. Arzneim-Forsch/Drug Res 1992; 42:238-246.
31. Gross SS, Wolin MS. Nitric oxide: pathophysiological mechanisms. Ann Rev Physiol 1995; 57:737-769.
32. Hagberg H, Andersson P, Lacarewicz J, Jacobson I, Butcher S, Sandberg M. Extracellular adenosine, inosine, hypoxanthine, and xanthine in relation to tissue nucleotides and purines in rat striatum during transient ischemia. J. Neurochem. 1987; 49:227-231.
33. Hansen AJ. Effect of anoxia on ion distribution in the brain. Physiol Rev 1985; 65:101-148.
34. Hawkins RA, Mans AM. Intermediary metabolism of carbohydrates and other fuels. In: Lajtha A ed. Handbook of neurochemistry. 2nd edition, vol. 3, Metabolism of the nervous system. New York: Plenum Press, 1983: 259-294.
35. Hille B. Ionic channels of excitable membranes. 3rd edition. Sunderland, Mass: Sinauer,2001.
36. Holmqvist BI, Ostholm T, Alm P, Ekstrom P. Nitric oxide synthase in the brain of a teleost. Neurosci Lett 1994; 171:205-208.
37. Huang PL, Dawson TM, Bredt DS, Snyder SH, Fishman MC. Targeted disruption of the neuronal nitric oxide synthase gene. Cell 1993; 75:1273-86.
38. Hulbert AJ, Else PL. Evolution of mammalian endothermic metabolism: mitochondrial activity and cell composition. Am J Physiol 1989; 256:R63-R69.
39. Hulbert AJ, Else PL. The cellular basis of endothermic metabolism: a role for "leaky" membranes? News Physiol Sci 1990; 5:25-28.
40. Hylland PH, Nilsson GE. Evidence that acetylcholine mediates increased cerebral blood flow velocity in crucian carp through a nitric oxide-dependent mechanism. J Cereb Blood Flow Metab 1995; 15:519-524.
41. Hylland P, Nilsson GE, Lutz PL. Time course of anoxia induced increase in cerebral blood flow rate in turtles: evidence for a role of adenosine. J Cereb Blood Flow Metab 1994; 14:877-881
42. Hylland PH, Nilsson GE, Lutz PL. Role of nitric oxide in the elevation of cerebral blood flow induced by acetylcholine and anoxia in the turtle. J Cereb Blood Flow Metab 1996 ; 16:290-295..
43. Iadecola C, Pelligrino DA, Moskowitz MA, Lassen NA. Nitric oxide sythase inhibition and cerebrovascular regulation. J Cereb Blood Flow Metab 1994; 14:175-192.
44. Irikura K, Huang PL, Ma J, Lee WS, Dalkara T, Fishman MC, Dawson TM, Snyder SH, Moskowitz MA. Cerebrovascular alterations in mice lacking neuronal nitric oxide synthase gene expression. Proc Natl Acad Sci USA 1995; 92:6823-6827.
45. Ishii K, Chang B, Kerwin JF Jr, Huang ZJ, Murad F. N^G-nitro-L-arginine: a potent inhibitor of endothelium-derived relaxing factor formation. Eur J Pharmacol 1990; 176:219-223.

46. Jessell TM, Kandell ER. Synaptic transmission: A bidirectional and self-modifiable form of cell-cell communication. Cell 1993; 10(Suppl):1-30.
47. Jerison HJ. Brain evolution and dinosaur brains. Am Naturalist 1969; 103:575-588.
48. Jiang C, Xia Y, Haddad GG. Role of ATP-sensitive K^+ channels during anoxia: major differences between rat (newborn and adult) and turtle neurons. J Physiol (Lond) 1992; 448:599-612.
49. Johnson JW, Ascher P. Glycine potentiates the NMDA response in cultured mouse brain neurons. Nature 1987; 325:529-531.
50. Johnston, GAR. GABAC receptors: relatively simple transmitter-gated ion channels. Trends Pharmacol Sci 1996; 17:319-323.
51. Jonas P, Koh DS, Kampe K, Hermsteiner M, Vogel W. ATP-sensitive and Ca-activated K channels in vertebrate axons: Novel links between metabolism and excitability. Pflügers Arch 1991; 418:68-73.
52. Kaila K. Ionic basis of $GABA_A$ receptor channel function in the nervous system. Prog Neurobiol 1994; 42:489-537.
53. Kandel ER, Schwartz JH, Jessell TM. Principles of Neural Science. 3rd ed. New York: Elsevier,1992.
54. Kaufman S. Regulatory properties of phenylalanine, tyrosine and tryptophan hydroxylases. Biochem Soc Trans 1985; 13:433-436.
55. Kelly RB. Storage and release of neurotransmitters. Cell 1993; 72:43-53.
56. Kennedy C, Sakurada O, Shinohara M, Miyaoka M. Local cerebral glucose utilization in the normal conscious macaque monkey. Ann Neurol 1978; 4:293-301.
57. Kirshner N. Biosynthesis of the catecholamines. In: Geiger SR ed. Handbook of Physiology. Sect. 7, Vol VI. Bethesda: American Physiological Society, 1975: 341-355.
58. Leone AM, Palmer RM, Knowles RG, Francis PL, Ashton DS, Moncada S. Constituitive and inducible nitric oxide synthases incorporate molecular oxygen into both nitric oxide and citrulline. J Biol Chem 1991; 266:23790-23795.
59. Lipton P, Whittingham TS. Energy metabolism and brain slice function. In: Dingledine R, ed. Brain Slices. New York: Plenum Press,1984: 113-153.
60. Lutz PL. Mechanisms for anoxic survival in the vertebrate brain. Ann Rev Physiol 1992; 54:601-618.
61. Magistretti PJ, Hof PR, Martin J-L. Adenosine stimulates glycogenolysis in mouse cerebral cortex: a possible coupling mechanism between neuronal activity and energy metabolism. J Neurosci 1986; 6:2558-2562.
62. Mason ST. Catecholamines and Behaviour. Cambridge, UK: Cambridge University Press,1984.
63. Mayer ML, Westbrook GL, Guthrie PB. Voltage-dependent block by Mg^{2+} of NMDA responses in spinal cord neurones. Nature 1984; 309:261-263.
64. McGeer PL, McGeer EG. Amino acid neurotransmitters. In: Siegel GJ, Agranoff B, Alberts RW (eds). Basic Neurochemistry: Molecular, Cellular, and Medical Aspects. 4th ed. New York: Raven Press,1989: 311-332.
65. Michenfelder JD. The interdependency of cerebral functional and metabolic effects following massive doses of thiopental in the dog. Anesthesiology 1974; 41:231-236.
66. Mink JW, Blumenschine RJ, Adams DB. Ratio of central nervous system to body metabolism in vertebrates: its constancy and functional basis. Am J Physiol 1981; 241:R203-R212.
67. Moller, P. Electric Fishes. London: Chapman and Hall, 1995.
68. Mongaghan DT, Cotman CW. Distribution of N-methyl-D-aspartate-sensitive L-glutamate binding sites in the rat brain. J Neurosci 1985; 5:2909-2917.
69. Mongaghan DT, Bridges RJ, Cotman CW. The excitatory amino acid receptors: their classes, pharmacology, and distinct properties in the function of the central nervous system. Ann Rev Pharmacol Toxicol 1989; 29:365-402.

70. Morii S, Ngai AC, Ko KR, Winn HR. Role of adenosine in regulation of cerebral blood flow: effects of theophylline during normoxia and hypoxia. Am J Physiol 1987; 253:H165-H175.
71. Mülsch A, Busse R N^G-nitro-L-arginine (N^5-[imino(nitroamino)methyl]-L-ornithine) impairs endothelium-dependent dilations by inhibiting cytosolic nitric oxide synthesis from L-arginine. Naunyn-Schmiedeberg's Arch Pharmacol 1990; 341:143-147.
72. Nagatsu T. Biochemistry of Catecholamines. Tokyo: University Park Press,1973.
73. Newby AC, Worku Y, Meghji P, Nakazawa, M, Skladanowski AC. Adenosine: a retaliatory metabolite or not? News Physiol Sci 1990; 5:67-70.
74. Nilsson GE. Neurotransmitters and anoxia resistance – comparative physiological and evolutionary perspectives. In: Surviving Hypoxia: Mechanisms of Control and Adaptation. P. W. Hochachka, P. L. Lutz, T. Sick, M. Rosenthal and G.van den Thilart (Eds). Boca Raton, FL: CRC Press, 1993: 401-413.
75. Nilsson GE. Brain and body oxygen requirements of *Gnathonemus petersii*, a fish with an exceptionally large brain. J Exp Bio 1996; 199:603-607.
76. Nilsson GE, Hylland P, Löfman CO. Anoxia and adenosine induce increased cerebral blood flow in crucian carp. Am J Physiol 1994; 267: R590-R595.
77. Nilsson GE, Lutz PL. Role of GABA in hypoxia tolerance, metabolic depression and hibernation – possible links to neurotransmitter evolution. Comp Biochem Physiol 1993; 105C:329-336.
78. Nilsson GE, Söderström V. Comparative aspects on nitric oxide in brain and its role as a cerebral vasodilator. Comp Biochem Physiol 1997; 118A:949-58.
79. Noma A. ATP-regulated K^+ channels in cardiac muscle. Nature 1983; 305:147-148.
80. O'Dell TJ, Huang PL, Dawson TM, Dinerman JL, Snyder SH, Kandel ER, Fishman MC. Endothelial NOS and the blockade of LTP by NOS inhibitors in mice lacking neuronal NOS. Science 1994; 265:542-546.
81. Olney JW. Neurotoxicity of NMDA receptor antagonists: an overview. Psychopharmacol Bull 1994; 30:533-540.
82. Oshima N. Adenosine inhibits the release of neurotransmitters from melanosome-aggregating nerves of fish. Comp Biochem Physiol C Comp Pharmacol 1989; 93:207-211.
83. Palmer RMJ, Ferrige AG, Moncada S. Nitric oxide accounts for the biological activity of endothelium-derived relaxing factor. Nature 1987; 327:524-526.
84. Phillis JW, O'Regan MH, Perkins LM. Adenosine 5'-triphosphate release from the normoxic and hypoxic in vivo rat cerebral cortex. Neurosci Lett 1993; 151:94-96.
85. Proctor WR, Dunwiddie TV. Pre- and postsynaptic actions of adenosine in the in vitro rat hippocampus. Brain Res 1987; 426:187-190.
86. Ransom RW, Stec NL. Cooperative modulation of MK-801 binding to the N-methyl-D-aspartate receptor-ion channel complex by L-glutamate, glycine, and polyamines. J Neurochem 1988; 51:830-836
87. Robin ED, Lewiston N, Newman A, Simon LM, Theodore J. Bioenergetic pattern of the turtle brain and resistance to profound loss of mitochondrial ATP generation. Proc Natl Acad Sci USA 1979; 76:3922-3926.
88. Rorsman P, Berggren PO, Bokvist K, Ericson H, Mohler H, Ostenson CG, Smith PA. Glucose-inhibition of glucagon secretion involves activation of $GABA_A$-receptor chloride channels. Nature 1989; 341:233-236.
89. Rosati AM, Traversa U, Lucchi R, Poli A. Biochemical and pharmacological evidence for the presence of A1 but not A2a adenosine receptors in the brain of the low vertebrate teleost *Carassius auratus* (goldfish). Neurochem Int 1995; 26:411-23.
90. Rothman SM, Olney JW. Excitotoxicity and the NMDA receptor – still lethal after eight years. Trends Neurosci 1995; 18:57-58.
91. Rudolphi KA, Schubert P, Parkinson FE, Fredholm BB. Adenosine and brain ischemia. Cerebrovasc. Brain Metab Rev 1992; 4:346-369.
92. Siesjö BK. Brain Energy Metabolism. Chichester: Wiley, 1978.
93. Siesjö BK. Calcium, excitotoxins, and brain damage. News Physiol Sci 1990; 5:120-125.

94. Sokoloff L, Reivich M, Kennedy C, Des Rosiers MH, Patlak CS, Pettigrew KD, Sakurada O, Shinohara M. The [^{14}C]deoxyglucose method for the measurement of local cerebral glucose utilization: theory, procedure, and normal values in the conscious and anaesthetized albino rat. J Neurochem 1977; 28:897-916.
95. Sommer B, Seeburg PH. Glutamate receptor channels: novel properties and new clones. Trends Pharmacol Sci 1992; 13:291-296.
96. Soubrié P. Reconciling the role of central serotonin neurons in human and animal behaviour. Behav Brain Sci 1986; 9:319-364
97. Southan GJ, Szabo C. Selective pharmacological inhibition of distinct nitric oxide synthase isoforms. Biochem Pharmacol 1996; 51:383-394.
98. Staley KJ, Soldo BL, Proctor WR. Ionic mechanisms of neuronal excitation by inhibitory GABAA receptors. Science 1995; 269:977-981.
99. Stiles GL. Adenosine receptors: physiological regulation and biochemical mechanisms. News Physiol Sci 1991; 6:161-165.
100. Stone TW. Physiological roles of adenosine and adenosine 5'-triphosphate in the nervous system. Neurosci 1981; 6:523-555.
101 Stone TW. Subtypes of NMDA receptors. Gen Pharmacol 1993; 24:825-832.
102. Suarez RK, Doll CJ, Buie AE, West TG, Funk GD, Hochachka PW. Turtles and rats: a biochemical comparison of anoxia-tolerant and anoxia-sensitive brains. Am J Physiol 1989; 26:R1083-R1088.
103. Sweeney MI. Neuroprotective effects of adenosine in cerebral ischemia: window of opportunity. Neurosci Biobehav Rev 1997; 21:207-17.
104. Toda N. Nitric oxide and the reulation of cerebral arterial tone. In: Vincent S, ed. Nitric Oxide in the Nervous System. London: Academic Press,1995.
105. Umans JG, Levi R. Nitric oxide in the regulation of blood flow and arterial pressure. Ann Rev Physiol 1995; 57:771-790.
106. Valtorta F, Meldolesi J, Fesce R. Synaptic vesicles: Is kissing a matter of competence? Trends Cell Biol 2001; 11:324-328.
107. Von Bonin G. Brain-weight and body-weight of mammals. J Gen Psychol 1937; 16:379-389.
108. Vincent SR. Nitric Oxide in the Nervous System. London: Academic Press, 1995.
109. White TD, Hoehn K. Release of adenosine and ATP from nervous tissue. In: Stone WT ed. Adenosine and the nervous system. London: London Academic Press,1991: 173-195.
110. Winberg S, Nilsson GE. Roles of brain monoamines in agonistic behaviour and stress reactions, with particular reference to fish. Comp Biochem Physiol 1993; 106C:597-614.
111. Wroblewski JT, Danysz W. Modulation of glutamate receptors: molecular mechanisms and functional implications. Ann. Rev. Pharmacol Toxicol 1989; 29:441-474.
112. Yarowsky P, Jehle J, Ingvar DH. Relationship between functional activity and glucose utilization in the rat superior cervical ganglion in vivo. Soc Neurosci Abstr 1979; 5:421.
113. Yoshino M, Obata T, Sho S, Kinemuchi H. Enzymatic and molecular characteristics of a new form of monoamine oxidase, distinct from form-A and form-B. Jpn J Pharmacol 1984; 35:105-115.
114. Zhang J, Snyder SH. Nitric oxide in the nervous system. Ann Rev Pharmacol Toxicol 1995; 35:213-233.

CHAPTER 3

OXYGEN SENSING AND HYPOXIA SIGNALING

1. INTRODUCTION

With the notable exception of specialized oxygen sensing cells in the carotid body, it was for a long time thought that the way most cells sense a fall in oxygen was via the downstream effects of hypoxia on ATP levels. Lately it has become clear that many, if not most cells can actually sense oxygen itself and thereby respond to hypoxia much earlier than allowed by the crisis of falling ATP, which would bring the cell very close to death before it could elicit a response. When mammals are subjected to hypoxia there is an increased expression of a number of physiologically important proteins involved in oxygen transport. For example a hypoxia induced increase in erythropoietin results in the enhanced formation of red blood cells which deliver oxygen to tissues. Hypoxia can also result in an elevated expression of vascular endothelial growth factor (VEGF) which causes an increase in capillaries in hypoxic tissues. The increased tissue vascularization enhances O_2 delivery providing protection against ischemic damage. In brain VEGF upregulation has been seen in transient global ischemia. Some cell types located in the carotid bodies, airway neuroepithelial bodies and pulmonary arteries are specialized oxygen-sensing cells involved in activating mechanisms to enhance oxygen delivery. The glomus cells of the carotid body, for example, up-regulate tyrosine hydroxylase which mediates the process of signaling to the central respiratory centers to enable the animal to attain a sustained increase in ventilation.

In a wide variety of cell types including specialized oxygen sensing cells, the response to hypoxia depends upon the existence of O_2 sensing

The Brain Without Oxygen. Third Edition. Peter L. Lutz, Göran E. Nilsson and Howard M. Prentice.

mechanisms that detect the fall in the partial pressure of O_2 and initiate a signaling pathway leading to physiologically appropriate changes in gene expression. The precise nature of these sensors is in most cases unknown, but several suggestions have been made. The sensors may be ubiquitously expressed membrane bound NADPH oxidases or alternatively components of the mitochondrial electron transport chain.[37,7] Other candidates include hypoxia sensitive ion channels which are found in pulmonary artery smooth muscle cells, CNS neurons, neuroepithelial cells and other cell types.[50,43] The signaling pathway involving the transcription factor HIF-1-alpha may itself contain the sensor in the form of an oxygen dependent enzyme that modifies the ability of HIF-1-alpha to interact with partners for ubiquitin mediated degradation.[31,32]

2. THEORIES OF OXYGEN SENSING

A long held view proposed that the oxygen sensor may be a heme protein.[24] According to this hypothesis hypoxia sensing processes that result in an induction of hypoxia induced gene expression occur through an allosteric shift in a protein that is capable of binding oxygen at a heme site. When human hepatoma cells were incubated with increasing concentrations of cobalt or other transition metals there was an induction of erythropoietin gene expression that was similar to the corresponding effect of hypoxia. Under these conditions it was proposed that the ferrous heme iron was replaced by other non-oxygen binding cations and thus the oxygen sensor would become locked into a deoxy-configuration.[24] Carbon monoxide was found to inhibit the hypoxic but not the cobalt induced erythropoietin expression, which was fully consistent with the inability of carbon monoxide to bind to cobalt or nickel substituted heme.[6] Additional data on iron chelators such as desferrioxamine demonstrating induced erythropoietin gene expression, has further added support to the heme protein hypothesis of oxygen sensing.[30] One problem, however, with the heme protein hypothesis, which makes it controversial, arises from data on inhibitors of heme synthesis which fail to extinguish the hypoxia responsive induction in gene expression.[64] In light of this data it would still be possible for a heme protein to be an oxygen sensor, assuming that cobalt acts on a component of a downstream pathway.

A number of studies have pointed to a particular heme molecule, the cytochrome b/NAD(P)H oxidase complex, as an oxygen sensor in mammalian cells.[37] While this molecule generates superoxide in the plasma membrane of phagocytes and B lymphocytes, the subunits of this complex are also found in other cell types including hepatoma cells, carotid body glomus cells, and pulmonary neuroepithelial bodies.[26,38,76] In neutrophils, this oxidase system comprises a catalytic complex of gp91phox and p22phox which forms a flavo-cytochrome b558 that is membrane bound, and cytosolic regulatory components including p47phox, p67phox and the GTP binding proteins Rac and Rap1A.[12] Expression of a number of these components has been demonstrated in oxygen sensing cells, and on this basis the cytochrome b558 complex was initially proposed to be a potential oxygen sensor. This complex is unlikely to be a universal oxygen sensor as studies on cell lines derived from patients with chronic granulomatous disease, where one subunit of the b558/NADPH complex is defective, demonstrate a pattern of normal oxygen-regulated gene expression.[73] Further evidence against this complex representing a universal oxygen sensor has been obtained from knockout mice deficient in the gp91-phox subunit of the b558/NADPH oxidase complex.[1] In these mice there was a decrease in the generation of ROS at the lung surface and yet no apparent abolition of the classical responses to hypoxia, including the lung vasoconstrictor response and whole cell K^+ current inhibition, in pulmonary artery smooth muscle cells. In the NAD(P)H oxidase model of oxygen sensing, electrons derived from NAD(P)H are carried to O_2, resulting in generation of superoxide and a more oxidized cytosolic redox state. Superoxide is rapidly converted to hydrogen peroxide (H_2O_2) by superoxide dismutase. H_2O_2 is in turn involved in local Fenton reactions at sites where ferrous iron is bound.[37] The Fenton reaction is a non- enzymatic conversion of hydrogen peroxide to a hydroxyl anion and a hydroxyl radical through oxidation of the ferrous iron. This reaction could cause a site specific oxidation of a regulatory protein, including a HIF-1-alpha family member. Indeed it has been proposed that under hypoxic conditions, HIF-1-alpha is released from suppression by a decrease in free radical concentrations.[74]

While a cytochrome b-type NAD(P)H oxidoreductase has been proposed as the oxygen sensor, its exact molecular identity remains to be determined. The neutrophil gp91phox is not involved in oxygen sensing.[73] Also the transcript for a nonphagocytic gp91phox like protein is not expressed in HEP3B cells where a cytochrome b-type NAD(P)H oxidoreductase has been proposed as the oxygen sensor.[79] To determine the molecular identity of a potential oxygen sensor candidate, Bunn and colleagues employed a genetic approach to clone a cDNA for a cytosolic cytochtome b-type NAD(P)H oxidoreductase – a flavohemoprotein (b5 + b5R) that contains cytochrome b5 and b5 reductase domains.[79] The two functional domains b5 and b5R are linked by an approximately 100 amino-acid long hinge region with a novel amino acid sequence. Interestingly the b5 + b5R protein is located in close proximity to the nucleus which could be consistent with a potential involvement in transduction of a hypoxia signal to a member of the HIF-1-alpha transcription factor family (see below), which would be in a similar location.

The mitochondrion has also been suggested as the oxygen sensor. Specifically Schumacker and colleagues have proposed that cytochrome c oxidase is implicated in this mitochondrial model of oxygen sensing.[7] Treatment of a variety of cell types with mitochondrial inhibitors including the complex I inhibitors, diphenyleneiodonium (DPI) and

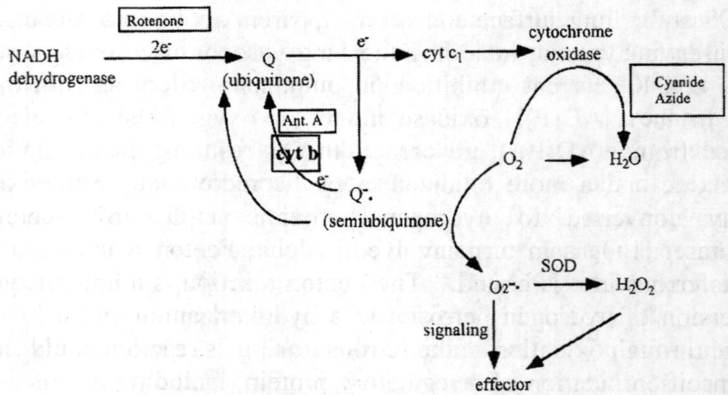

Fig. 3.1 The mitochondrial hypothesis of hypoxia sensing. This figure illustrates the effect of hypoxia on the mitochondrial electron transport chain. In hypoxia there is an inhibition of the reduction of O_2 to H_2O by cytochrome oxidase. This results in release of electrons upstream at the ubisemiquinone cycle (complex III) and generation of superoxide.(after Lopez-Barneo et al., 2001).

rotenone, have been found to result in abolition of hypoxic responses. Hepatoma cells depleted of mitochondrial DNA, known as rho 0 cells, demonstrate a block in hypoxic induction of oxygen regulated genes.[8] However inhibitors of cytochrome c oxidase, such as cyanide and azide failed to interfere with the hypoxia responses suggesting that mitochondria function as oxygen sensors in some cell types or experimental systems but not in others.[64]

The role of reactive oxygen species in hypoxia sensing has become controversial. The NAD(P)H oxidase model of oxygen sensing would predict that, in hypoxia, a decrease in ROS would signal the cellular oxygenation state. In the mitochondrial model, reduction of O_2 to H_2O by cytochrome oxidase (complex IV) is inhibited in hypoxia with a release of electrons upstream at complex III and generation of superoxide. It has been difficult to distinguish between the NAD(P)H and mitochondrial models of O_2 sensing because of technical limitations in measurements on reactive oxygen species concentrations.[59]

3. HYPOXIA SIGNALING IN BRAIN

While the specific nature of the oxygen sensor(s) remains unclear the components of some of the principal hypoxia signal transduction pathways have been well characterized. In particular hypoxic up-regulation of a number of physiologically relevant genes, including those that encode erythropoietin, VEGF, and a number of glycolytic enzymes, is under the control of the HIF family of transcription factors. HIF-1 was originally isolated as a factor that binds to the 3' flanking region of the erythropoietin gene.[71]

Hypoxia inducible HIF-1 expression and DNA binding activity have been reported in a variety of cell lines[69,70] and in different mouse, rat and ferret organs.[75,78] HIF-1 and HIF-1 mRNA are expressed in brain tissue and mediate induction of genes for maintaining oxygen homeostasis and neuroprotection. Hypoxic induction of HIF-1 has been observed in neurons in culture and in brain tissue in vivo. In focal brain ischemia in rat for example, HIF-1-alpha and its target genes are induced in the peri-infarct penumbra. In a mouse middle cerebral artery occlusion model, HIF-1 responsive reporter gene expression was demonstrated in ischemic brain following transduction with a Herpes Simplex Virus vector.[29]

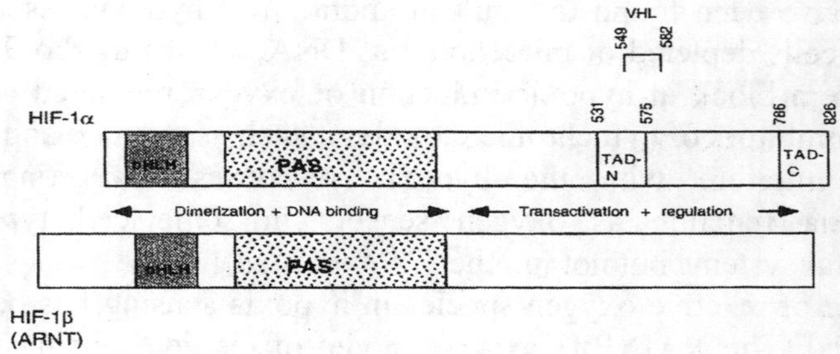

Fig. 3.2 Structure of HIF-1-alpha. The transactivation domain-N (TAD-N) overlaps with the oxygen dependent degradation domain (ODD) where pVHL binds to HIF-1-alpha under normoxic conditions. Gene activation under hypoxic conditions arises through dimerization of HIF-1-alpha and HIF-1-beta through interactions of their bHLH and PAS domains and subsequent translocation of the dimer to the nucleus. Further modifications of HIF-1 function may occur through reduction of a cysteine residue in the TAD-C domain enabling interaction with the cofactor p300 (after Semenza, 2001[59]) ischemia in rat for example, HIF-1-alpha and its target genes are induced in the peri-infarct penumbra. In a mouse middle cerebral artery occlusion model, HIF-1 responsive reporter gene expression was demonstrated in ischemic brain following transduction with a Herpes Simplex Virus vector.[29]

HIF-1 consists of a heterodimer between two sub-units HIF-1-alpha and HIF-1-beta. HIF-1-beta, also known as ARNT (aryl hydrocardbon nuclear translocator), is xpressed independently of oxygen concentration. By contrast, HIF-1-alpha is rapidly degraded under normoxia. This degradation is inhibited by hypoxia, and HIF-1-alpha is translocated to the nucleus and transactivates target genes after dimerisation with HIF-1-beta.[58] The dimerisation of HIF-1-alpha with HIF-1 beta is essential for DNA binding and promoter activation, but it is not required for HIF-1-alpha translocation to the nucleus.[22,27,10] The amino 390 amino acid domain of HIF-1 alpha is responsible for dimerisation with HIF-1-beta as well as for DNA binding.[33] In non-hypoxic cells HIF-1-alpha protein becomes primed for ubiquitination, a process involving covalent ligation to ubiquitin, a highly conserved small protein. In normoxia, signals acting via the HIF-1-alpha oxygen dependent degradation domain subsequently lead to degradation of the molecule in proteasomes. Under normoxic conditions HIF-1-alpha associates with the von-Hippel Landau tumor suppressor protein (pVHL) which constitutes the recognition component of an E3 ubiquitin-protein ligase complex that promotes ubiquitin dependent proteolysis of HIF-1-alpha.[45] VHL protein may be differentially expressed between various cell types and in brain development for example, increased VHL protein levels have been shown to correlate with neuronal differentiation.[36] By contrast, in hypoxic cells HIF-1-alpha degradation is inhibited resulting in increased dimerisation with HIF-1-beta and subsequent activation of inducible gene expression. It has recently been demonstrated that hypoxia not only regulates and prevents the interaction between HIF-1-alpha and pVHL, thus stabilizing HIF-1-alpha, but also induces the nuclear translocation of HIF-1-alpha. The key enzyme controlling the pVHL/HIF-1-alpha interaction is a specific HIF-1-alpha proline hydroxylase.[31,32] Proline hydroxylases attach an oxygen atom such as a hydroxyl group, to the C-4 of proline. This reaction requires the iron atom in the reduced ferrous state, which is consistent with the need for a reducing agent such as ascorbic acid (see chapter 7).

4. MECHANISMS OF TRANSCRIPTIONAL ACTIVATION BY HIF-1

Activation of gene expression by HIF-1 in a variety of tissues including brain and heart occurs through the binding of the HIF-1 complex to a core consensus site RCGTG present in the hypoxia response elements of target genes.[61] HIF-1 transactivates through the interactions of coactivators including CBP/p300, SRC-1 and TIF-2 with the transactivation domains of HIF-1-alpha and a transactivation domain of ARNT.[33,55] CBP/p300 is a histone acetyltransferase that is known to interact with a range of key transcription factors including p53, an apoptosis pathway component. There are two transactivation domains in HIF-1-alpha, one of which overlaps with the oxygen-dependent degradation domain (ODD).[35] The contributions of the transactivation domains might be further enhanced by the action of kinases such as p42/p44 MAPK (also known as the ERK2 and ERK1 respectively) or by redox modification.[56] In the nucleus this increased activation may be mediated through a redox -sensitive conserved cysteine residue in HIF-1-alpha.[16]

The mechanisms underlying the stabilization of the ODD has recently been discovered. This stabilization does not depend on p42/p44MAP kinase mediated phosphorylation nor on redox modification of the ODD residues. Instead stabilization occurs through hydroxylation of a single proline residue, p564, in the ODD domain of HIF-1-alpha.[31,32] The enzyme responsible for the hydroxylation of this key proline is the specific proline hydroxylase known as HIF-PH (see above). Remarkably a requirement for dioxygen as a cosubstrate and iron as a cofactor suggests that HIF-PH acts directly as an oxygen sensor.[31,32]

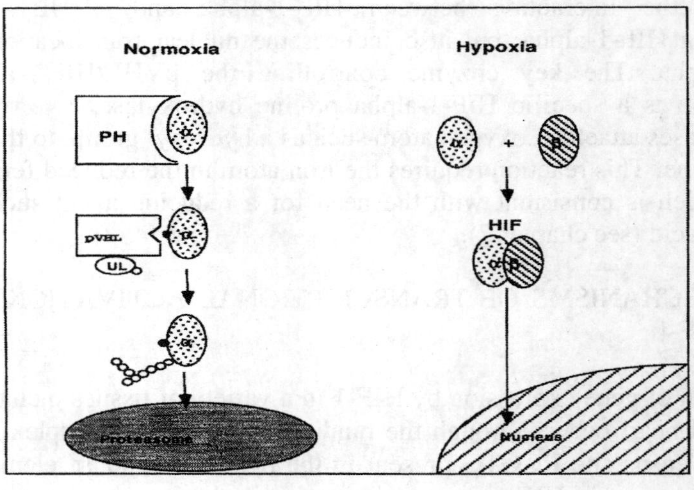

Fig. 3.3 The role of prolyl hydroxylase in HIF transcription factor regulation.
The interaction between pVHL and HIF-1-alpha is regulated by an iron and oxygen dependent proline hydroxylase that under normoxic conditions oxidatively modifies HIF-1-alpha enabling the interaction between HIF and pVHL and resulting in HIF-1-alpha degradation. In hypoxia HIF-1-alpha forms a heterodimer with HIF-1-beta and then translocates to the nucleus.

5. HYPOXIA RESPONSIVE GENES AND NEUROPROTECTION

In rat brain, middle cerebral artery occlusion induces HIF-1-alpha mRNA and causes an increase in both HIF-1-alpha and HIF-1-beta proteins. The mRNAs for HIF-1-alpha the HIF-1 target genes encoding glucose transporter-1, phosphofructokinase-L (PFK-L), aldose-A (ALD-A), pyruvate kinase-M (PK-M) and lactate dehydrogenase-A (LDH-A) were elevated at 7.5 h after initiation of ischemia with further increases at 19 and 24h.[3] In the core of the infarct, the oxygen and energy depletion occurred so rapidly that it precluded a HIF-1 mediated response. In the penumbra, however, a moderate and prolonged decrease in blood flow is effective at increasing HIF-1-alpha mRNA and a number of HIF-1 target genes.[3] The most thoroughly studied gene induced by hypoxia is the erythropoietin gene whose product is the principal regulator of erythropoiesis. Erythropoietin is best known as a factor produced in fetal liver and adult kidney in response to hypoxia. With recent reports that erythropoietin mRNA is expressed in rodent and primate brains, potential roles for the factor in neuronal survival have begun to be established.[44,65] Erythropoietin mRNA expression has been found in neuronal cell lines as well as hippocampal and cortical neuronal primary cultures, and erythropoietin mRNA is expresses in primary cultures of cortical astrocytes. Both astrocytic and neuronal expression of erythropoietin mRNA are induced by hypoxia in a process that is dependent on protein synthesis[4]. In adult brain erythropoietin mRNA expression is also hypoxia inducible. Induction of erythropoietin mRNA expression in a variety of cell types arises through HIF-1 signaling. It is significant that induction of erythropoietin gene expression via HIF-1 arises in response not only to hypoxia but also to hypoglycemia or to oxidative stress which are additional characteristics of ischemic tissues.[8,17]

Hypoxic exposure of a variety of cultured cell types and tissues results in increased VEGF levels. Increased VEGF protein expression can result from transcriptional regulation and from increased mRNA stability. Induction of VEGF expression has also been reported under conditions of normoxia with hypoglycemia.[57] In cerebral ischemia, where both oxygen and glucose are at low levels, as might have been anticipated, an induction of VEGF expression was observed.[52] Similarly reperfusion of ischemic tissue results in release of reactive oxygen intermediates which are also known to be inducers of VEGF expression.[39] VEGF is known to be an angiogenic peptide which is produced at high levels under hypoxic conditions and acts on endothelial cells to promote new blood vessel formation.[47,51] Recent studies have also demonstrated a neuroprotective effect of VEGF with increased survival of mouse superior cervical and dorsal root neurons, rat mesencephalic neurons and HN33 hippocampal cells.[63,62] In cultured HN33 cells subjected to hypoxia and glucose deprivation, VEGF promotes cell survival through VEGFR-2 receptors and the phosphatidylinositol 3'-kinase (PI3-K)/Akt signaling pathway.[34] In that study the HN33 cells did not show elevated expression of VEGF, which may be related to the severity of the hypoxic /hypoglycemic insult. It is known, for example, that VEGF is induced in neurons of the ischemic border zone but not in those of the ischemic core in focal cerebral ischemia.[40] VEGF induction in response to hypoxia has been demonstrated in astroglia but may also occur in neurons.[34]

6. HYPOXIA RESPONSE PATHWAYS THAT DO NOT INVOLVE HIF-1

Although HIF-1 has been found to mediate activation of a variety of hypoxia induced genes there are some notable exceptions amongst the muscle specific glycolytic enzyme genes. In particular while active HIF-1 sites have been confirmed in the promoter regions of the liver phosphoglycerate kinase (PGK-1), aldolase-A (ALD-A), enolase (Eno-1) and lactate dehydrogenase (LDH-A) genes, it has been shown that there are no such sites in the proximal promoter regions of either the muscle specific pyruvate kinase (PKM) gene or another muscle specific gene, the beta-enolase (beta-Eno) gene. Both of these genes are hypoxia inducible. The mechanism for hypoxia inducibility of the muscle specific PKM and beta-enolase promoters has been found to involve a zinc finger transcription factor Sp3, acting as a repressor in aerobic conditions and undergoing degradation and replacement with functional Sp1 under hypoxic conditions.[72] A second category of non-HIF-1 hypoxic response has been recently reported. It involves the transcriptional upregulation of the gene for Apoptosis Inhibitory Protein-2 (IAP-2) under conditions of severe hypoxia.[15] While this is an intriguing observation it has not yet been determined which regulatory elements of the IAP-2 gene are responsible for the induction by severe hypoxia.

7. OXYGEN SENSING ION CHANNELS

There is strong evidence that one set of acute cellular effects of hypoxia involves the inactivation of specific ion channels in an apparently direct and sensitive manner. These cellular effects of hypoxia have been particularly well studied in specific cell types that have been regarded as hypoxia sensing cells. Cells in the carotid body, for example, which is the principal arterial chemoreceptor, are known to mediate hypoxic signaling to maintain oxygen homeostasis in the body. Under conditions of low arterial pO_2, clusters of cells in the carotid body known as glomus cells respond by increasing afferent sensory information to the central respiratory centers.[25,42] Similar clusters of innervated neurosecretory cells are found in the neuroepithelial bodies where they respond to decreased pO_2 levels in air during inhalation. Neuroepithelial bodies are found near the airway bifurcation which is an ideal location for detecting changes in airway gas composition. While the exact nature of neurotransmitter release from these cells is unknown, the neuroepithelial body displays a number of features that identify it as a potential airway chemoreceptor. 5HT produced by the neuroepithelial body may play a role in bronchoconstriction or vasomotor tone or act as a growth factor.[77,19]

In the vasculature, hypoxic pulmonary vasoconstriction is manifested in pulmonary arteries and veins, particularly in the small resistance arteries where smooth muscle cells mediate this acute response.[53,48] In coronary and cerebral vessels, hypoxic vasodilation, mediated by K_{ATP} channels in the vascular smooth muscle, is critical for maintaining adequate O_2 delivery to heart and brain respectively.[14, 43]

Chemosensory transduction in glomus cells, and in other chemoreceptor cells, is initiated by a hypoxia regulated closure of K^+ channels leading to subsequent depolarization, Ca^{2+} influx, and transmitter release.[76,66] Oxygen sensitive K^+ channels were first detected in rabbit carotid body glomus or type 1 cells.[42] These K^+ channels were marked by the rapidity of their response to changes in pO_2 and the fact that hypoxia elicited the inhibition in excised membrane patches.[21] This latter fact revealed that the channels were hypoxia sensitive and were not responding to general changes in cellular metabolic status.[43] Lopez-

Barneo et al., have established that voltage gated Ca^{2+} channels can also be inhibited by hypoxia in glomus cells. This inhibition was voltage dependent and only seen in lower activating test potentials.[46] In hypoxic pulmonary vasoconstriction one or more membrane K^+ channels are inhibited leading to membrane depolarization, opening of voltage gated Ca^{2+} channels, release of Ca^{2+} from ryanodine sensitive stores and smooth muscle cell contraction.[67,53]

In coronary and cerebral vessels exposed to hypoxia, vasodilatation is not only mediated by K_{ATP} channels, but as mentioned above, there are influences from other membrane ion channels. This includes the closure of Ca^{2+} sensitive K^+ channels and L-type Ca^{2+} channels.[23,18]

Since the initial observation of O_2 sensitive K^+ channels in glomus cells a number of recombinant voltage gated potassium channel (Kv channel) subtypes have been implicated in hypoxia sensing in pulmonary artery smooth muscle cells. These voltage gated K^+ channels include the Shaker-like channels Kv1.2 and Kv1.5, and the recently investigated Kv3.1. The vast bulk of the research on hypoxia sensitive ion channels has been in non-neuronal cells and thus the brain remains largely unexplored in this regard. There is however evidence of a role for hypoxia regulated ion channels in brain, and in particular K^+ channel inhibition has been reported in excised membrane patches of central neurons. In cortical and hippocampal neurons, hypoxia induces hyperpolarization through activation of Ca^{2+} sensitive K^+ channels and in human neocortical neurons in tissue slices reduction of the amplitude of Na^+ currents by hypoxia or metabolic intoxication has been shown.[13,68]

8. SUMMARY

In summary the nature of the hypoxia sensor is still controversial, but it is possible that a number of cellular processes, including the mitochondrial electron transport chain and NAD(P)H oxidoreductases, may signal to modulate the activity of HIF-1-alpha or certain other non-HIF-1 related transcriptional activators. The latter could include Sp3 or the unknown activator of the IAP-2 gene. The recent discovery of a key role for a proline hydroxylase in HIF-1 stabilization now raises the possibility that the HIF-1 complex itself may act as one of the oxygen sensors in the cell. In specialized oxygen sensing cells as well as other cell types including central neurons, specific classes of membrane ion channels are directly regulated by hypoxia. The molecular identities have been established for specific voltage gated potassium channels that are regulated by hypoxia and these include Kv1.2, Kv1.5 and Kv3.1. Expression of voltage gated calcium channels, sodium channels and Ca^{2+} sensitive K^+ channels have also been shown to be modulated by hypoxia in different model systems and play a role in the process of hypoxia sensing.

9. REFERENCES

1. Archer SL, Reeve HL, Michelakis E, Puttagunta L, Waite R, Nelson DP, Dinauer MC, Weir EK.O_2 sensing is preserved in mice lacking the gp91 phox subunit of NADPH oxidase. Proc Natl Acad Sci U S A. 1999; 96(14):7944-9.
2. Archer SL, Weir EK, Reeve HL, Michelakis E. Molecular identification of O_2 sensors and O_2-sensitive potassium channels in the pulmonary circulation. Adv Exp Med Biol. 2000;475:219-40.
3. Bergeron M, Yu AY, Solway KE, Semenza GL, Sharp FR. Induction of hypoxia-inducible factor-1 (HIF-1) and its target genes following focal ischaemia in rat brain. Eur J Neurosci. 1999 (12):4159-70.
4. Bernaudin M, Bellail A, Marti HH, Yvon A, Vivien D, Duchatelle I, Mackenzie ET, Petit E. Neurons and astrocytes express EPO mRNA: oxygen-sensing mechanisms that involve the redox-state of the brain. Glia. 2000 (3):271-8.
5. Bunn HF, Gu J, Huang LE, Park JW, Zhu H. Erythropoietin: a model system for studying oxygen-dependent gene regulation. J Exp Biol. 1998; 201 (8):1197-201.
6. Bunn HF, Poyton RO. Oxygen sensing and molecular adaptation to hypoxia. Physiol Rev. 1996; 76(3):839-85.
7. Chandel NS, Schumacker PT.Cellular oxygen sensing by mitochondria: old questions, new insight. J Appl Physiol. 2000 ;88(5):1880-9.
8. Chandel NS, Maltepe E, Goldwasser E, Mathieu CE, Simon MC, Schumacker PT. Mitochondrial reactive oxygen species trigger hypoxia-induced transcription. Proc Natl Acad Sci U S A. 1998 ;95(20):11715-20.
9. Chandel NS, McClintock DS, Feliciano CE, Wood TM, Melendez JA, Rodriguez AM, Schumacker PT. Reactive oxygen species generated at mitochondrial complex III stabilize hypoxia-inducible factor-1alpha during hypoxia: a mechanism of O2 sensing. J Biol Chem. 2000; 275(33):25130-8.
10. Chilov D, Camenisch G, Kvietikova I, Ziegler U, Gassmann M, Wenger RH. Induction and nuclear translocation of hypoxia-inducible factor-1 (HIF-1): heterodimerization with ARNT is not necessary for nuclear accumulation of HIF-1alpha. J Cell Sci. 1999; 112 (8):1203-12.
11. Cockman ME, Masson N, Mole DR, Jaakkola P, Chang GW, Clifford SC, Maher ER, Pugh CW, Ratcliffe PJ, Maxwell PH. Hypoxia inducible factor-alpha binding and ubiquitylation by the von Hippel-Lindau tumor suppressor protein. J Biol Chem. 2000; 275(33):25733-41.
12. Cross AR, Erickson RW, Curnutte JT.The mechanism of activation of NADPH oxidase in the cell-free system: the activation process is primarily catalytic and not through the formation of a stoichiometric complex. 1: Biochem J 1999;341 (2):251-5.
13. Cummins TR, Jiang C, Haddad G.G. Human neocortical excitability is decreased during anoxia via sodium channel modulation. J Clin Invest. 1993; 91(2):608-15.
14. Dart C, Standen NB. Activation of ATP-dependent K+ channels by hypoxia in smooth muscle cells isolated from the pig coronary artery. J Physiol. 1995; 483 (1):29-39.
15. Dong Z, Venkatachalam MA, Wang J, Patel Y, Saikumar P, Semenza GL, Force T, Nishiyama J. Up-regulation of apoptosis inhibitory protein IAP-2 by hypoxia. Hif-1-independent mechanisms. J Biol Chem. 2001; 276(22):18702-9.

16. Ema M, Hirota K, Mimura J, Abe H, Yodoi J, Sogawa K, Poellinger L, Fujii-Kuriyama Y. Molecular mechanisms of transcription activation by HLF and HIF1alpha in response to hypoxia: their stabilization and redox signal-induced interaction with CBP/p300. EMBO J. 1999; 18(7):1905-14.
17. Fandrey J, Frede S, Ehleben W, Porwol T, Acker H, Jelkmann W. Cobalt chloride and desferrioxamine antagonize the inhibition of erythropoietin production by reactive oxygen species. Kidney Int. 1997; 51(2):492-6.
18. Franco-Obregon A, Lopez-Barneo J. Differential oxygen sensitivity of calcium channels in rabbit smooth muscle cells of conduit and resistance pulmonary arteries. J Physiol. 1996; 491 (2):511-8.
19. Fu XW, Nurse CA, Wang YT, Cutz E. Selective modulation of membrane currents by hypoxia in intact airway chemoreceptors from neonatal rabbit. J Physiol. 1999; 514 (1):139-50.
20. Fu XW, Wang D, Nurse CA, Dinauer MC, Cutz E. NADPH oxidase is an O2 sensor in airway chemoreceptors: evidence from K+ current modulation in wild-type and oxidase-deficient mice. Proc Natl Acad Sci U S A. 2000; 97(8):4374-9.
21. Ganfornina MD, Lopez-Barneo J. Single K+ channels in membrane patches of arterial chemoreceptor cells are modulated by O2 tension. Proc Natl Acad Sci U S A. 1991; 88(7):2927-30.
22. Gassmann M, Kvietikova I, Rolfs A, Wenger RH. Oxygen- and dioxin-regulated gene expression in mouse hepatoma cells. Kidney Int. 1997; 51(2):567-74.
23. Gebremedhin D, Bonnet P, Greene AS, England SK, Rusch NJ, Lombard JH, Harder DR. Hypoxia increases the activity of Ca(2+)-sensitive K+ channels in cat cerebral arterial muscle cell membranes. Pflugers Arch. 1994; 428(5-6):621-30.
24. Goldberg MA, Dunning SP, Bunn HF. Regulation of the erythropoietin gene: evidence that the oxygen sensor is a heme protein. Science 1988; 242(4884):1412-5
25. Gonzalez C, Almaraz L, Obeso A, Rigual R. Carotid body chemoreceptors: from natural stimuli to sensory discharges. Physiol Rev. 1994; 74(4):829-98.
26. Gorlach A, Fandrey J, Holtermann G, Acker H Effects of cobalt on haem proteins of erythropoietin-producing HepG2 cells in multicellular spheroid culture. FEBS Lett 1994; 348(2):216-8
27. Gradin K, McGuire J, Wenger RH, Kvietikova I, Whitelaw ML, Toftgard R, Tora L, Gassmann M, Poellinger L.Functional interference between hypoxia and dioxin signal transduction pathways: competition for recruitment of the Arnt transcription factor. Mol Cell Biol. 1996; 16(10):5221-31.
28. Haddad JJ, Olver RE, Land SC. Antioxidant/pro-oxidant equilibrium regulates HIF-1alpha and NF-kappa B redox sensitivity. Evidence for inhibition by glutathione oxidation in alveolar epithelial cells. J Biol Chem. 2000; 275(28):21130-9.
29. Halterman MW, Federoff HJ. HIF-1alpha and p53 promote hypoxia-induced delayed neuronal death in models of CNS ischemia. Exp Neurol. 1999; 159(1):65-72.
30. Ho VT, Bunn HF. Effects of transition metals on the expression of the erythropoietin gene: further evidence that the oxygen sensor is a heme protein. Biochem Biophys Res Commun 1996; 223(1):175-80
31. Ivan M, Kondo K, Yang H, Kim W, Valiando J, Ohh M, Salic A, Asara JM, Lane WS, Kaelin WG Jr. HIFalpha targeted for VHL-mediated destruction by proline hydroxylation: implications for O2 sensing. Science. 2001 ;292(5516):464-8.
32. Jaakkola P, Mole DR, Tian YM, Wilson MI, Gielbert J, Gaskell SJ, Kriegsheim Av, Hebestreit HF, Mukherji M, Schofield CJ, Maxwell PH, Pugh CW, Ratcliffe PJ. Targeting of HIF-alpha to the von Hippel-Lindau ubiquitylation complex by O_2-regulated prolyl hydroxylation. Science. 2001; 292(5516):468-72.
33. Jiang BH, Zheng JZ, Leung SW, Roe R, Semenza GL. Transactivation and inhibitory domains of hypoxia-inducible factor 1alpha. Modulation of transcriptional activity by oxygen tension. J Biol Chem. 1997; 272(31):19253-60.
34. Jin KL, Mao XO, Greenberg DA. Vascular endothelial growth factor: direct neuroprotective effect in in vitro ischemia. Proc Natl Acad Sci U S A. 2000; 97(18):10242-7.

35. Kamura T, Sato S, Iwai K, Czyzyk-Krzeska M, Conaway RC, Conaway JW Activation of HIF1alpha ubiquitination by a reconstituted von Hippel-Lindau (VHL) tumor suppressor complex. Proc Natl Acad Sci U S A. 2000; 97(19):10430-5.
36. Kanno H, Saljooque F, Yamamoto I, Hattori S, Yao M, Shuin T, U HS. Role of the von Hippel-Lindau tumor suppressor protein during neuronal differentiation. Cancer Res. 2000; 60(11):2820-4.
37. Kietzmann T, Porwol T, Zierold K, Jungermann K, Acker H. Involvement of a local fenton reaction in the reciprocal modulation by O_2 of the glucagon-dependent activation of the phosphoenolpyruvate carboxykinase gene and the insulin-dependent activation of the glucokinase gene in rat hepatocytes. Biochem J. 1998; 335 (Pt 2):425-32.
38. Kummer W, Acker H. Immunohistochemical demonstration of four subunits of neutrophil NAD(P)H oxidase in type I cells of carotid body. : J Appl Physiol 1995; 78(5):1904-9
39. Kuroki M, Voest EE, Amano S, Beerepoot LV, Takashima S, Tolentino M, Kim RY, Rohan RM, Colby KA, Yeo KT, Adamis AP. Reactive oxygen intermediates increase vascular endothelial growth factor expression in vitro and in vivo. J Clin Invest. 1996; 98(7):1667-75.
40. Lennmyr F, Ata KA, Funa K, Olsson Y, Terent A. Expression of vascular endothelial growth factor (VEGF) and its receptors (Flt-1 and Flk-1) following permanent and transient occlusion of the middle cerebral artery in the rat. J Neuropathol Exp Neurol. 1998; 57(9):874-82.
41. Levitan ES, Gealy R, Trimmer JS, Takimoto K. Membrane depolarization inhibits Kv1.5 voltage-gated K+ channel gene transcription and protein expression in pituitary cells. J Biol Chem. 1995; 270(11):6036-41.
42. Lopez-Barneo J, Lopez-Lopez JR, Urena J, Gonzalez C. Chemotransduction in the carotid body: K+ current modulated by PO2 in type I chemoreceptor cells. Science. 1988; 241(4865):580-2.
43. Lopez-Barneo J, Pardal R, Ortega-Saenz P. Cellular mechanism of oxygen sensing. Annu Rev Physiol. 2001;63:259-87.
44. Marti HH, Wenger RH, Rivas LA, Straumann U, Digicaylioglu M, Henn V, Yonekawa Y, Bauer C, Gassmann M. Erythropoietin gene expression in human, monkey and murine brain. Eur J Neurosci. 1996; 8(4):666-76.
45. Maxwell PH, Wiesener MS, Chang GW, Clifford SC, Vaux EC, Cockman ME, Wykoff CC, Pugh CW, Maher ER, Ratcliffe PJ. The tumour suppressor protein VHL targets hypoxia-inducible factors for oxygen-dependent proteolysis. Nature. 1999; 399(6733):271-5.
46. Montoro RJ, Urena J, Fernandez-Chacon R, Alvarez de Toledo G, Lopez-Barneo J. Oxygen sensing by ion channels and chemotransduction in single glomus cells. J Gen Physiol. 1996; 107(1):133-143.
47. Neufeld G, Cohen T, Gengrinovitch S, Poltorak Z. Vascular endothelial growth factor (VEGF) and its receptors. FASEB J. 1999 Jan;13(1):9-22.
48. Osipenko ON, Evans AM, Gurney AM. Regulation of the resting potential of rabbit pulmonary artery myocytes by a low threshold, O2-sensing potassium current. Br J Pharmacol. 1997; 120(8):1461-70.
49. Peers C. Oxygen-sensitive ion channels. Trends Pharmacol Sci. 1997; 18(11):405-8.
50. Perez-Garcia MT, Lopez-Lopez JR. Are Kv channels the essence of O2 sensing? Circ Res. 2000; 86(5):534-40
51. Petrova TV, Makinen T, Alitalo K. Signaling via vascular endothelial growth factor receptors. Exp Cell Res. 1999; 253(1):117-30.
52. Pichiule P, Chavez JC, Xu K, LaManna JC. Vascular endothelial growth factor upregulation in transient global ischemia induced by cardiac arrest and resuscitation in rat brain. Brain Res Mol Brain Res. 1999; 74(1-2):83-90.
53. Post JM, Gelband CH, Hume JR. [Ca2+]i inhibition of K+ channels in canine pulmonary artery. Novel mechanism for hypoxia-induced membrane depolarization. Circ Res. 1995; 77(1):131-9.
54. Post JM, Hume JR, Archer SL, Weir EK. Direct role for potassium channel inhibition in hypoxic pulmonary vasoconstriction. Am J Physiol. 1992; 262(4):C882-90.

55. Pugh CW, O'Rourke JF, Nagao M, Gleadle JM, Ratcliffe PJ. Activation of hypoxia-inducible factor-1; definition of regulatory domains within the alpha subunit. J Biol Chem. 1997; 272(17):11205-14.
56. Richard DE, Berra E, Gothie E, Roux D, Pouyssegur J. p42/p44 mitogen-activated protein kinases phosphorylate hypoxia-inducible factor 1alpha (HIF-1alpha) and enhance the transcriptional activity of HIF-1. J Biol Chem. 1999; 274(46):32631-7.
57. Satake S, Kuzuya M, Miura H, Asai T, Ramos MA, Muraguchi M, Ohmoto Y, Iguchi A. Up-regulation of vascular endothelial growth factor in response to glucose deprivation. Biol Cell. 1998; 90(2):161-8.
58. Semenza GL. Chairman's summary: mechanisms of oxygen homeostasis, circa 1999. Adv Exp Med Biol. 2000;475:303-10.
59. Semenza GL HIF-1 and mechanisms of hypoxia sensing. Curr Opin Cell Biol. 2001; 13(2):167-71.
60. Semenza GL. Hypoxia-inducible factor 1: master regulator of O2 homeostasis. Curr Opin Genet Dev. 1998 ;8(5):588-94.
61. Semenza GL. Hypoxia-inducible factor 1: control of oxygen homeostasis in health and disease. Pediatr Res. 2001; 49(5):614-7.
62. Silverman WF, Krum JM, Mani N, Rosenstein JM. Vascular, glial and neuronal effects of vascular endothelial growth factor in mesencephalic explant cultures. Neuroscience. 1999;90(4):1529-41.
63. Sondell M, Lundborg G, Kanje M. Vascular endothelial growth factor has neurotrophic activity and stimulates axonal outgrowth, enhancing cell survival and Schwann cell proliferation in the peripheral nervous system. J Neurosci. 1999; 19(14):5731-40.
64. Srinivas V, Leshchinsky I, Sang N, King MP, Minchenko A, Caro J. Oxygen sensing and HIF-1 activation does not require an active mitochondrial respiratory chain electron-transfer pathway. J Biol Chem. 2001; 276(25):21995-8.
65. Tan CC, Eckardt KU, Firth JD, Ratcliffe PJ. Feedback modulation of renal and hepatic erythropoietin mRNA in response to graded anemia and hypoxia. Am J Physiol. 1992; 263(3 Pt 2):F474-81.
66. Thompson RJ, Jackson A, Nurse CA. Developmental loss of hypoxic chemosensitivity in rat adrenomedullary chromaffin cells. J Physiol. 1997; 498 (Pt 2):503-10.
67. Vadula MS, Kleinman JG, Madden JA. Effect of hypoxia and norepinephrine on cytoplasmic free Ca2+ in pulmonary and cerebral arterial myocytes. Am J Physiol. 1993; 265(6 Pt 1):L591-7.
68. Wang D, Youngson C, Wong V, Yeger H, Dinauer MC, Vega-Saenz Miera E, Rudy B, Cutz E. NADPH-oxidase and a hydrogen peroxide-sensitive K+ channel may function as an oxygen sensor complex in airway chemoreceptors and small cell lung carcinoma cell lines. Proc Natl Acad Sci U S A. 1996; 93(23):13182-7.
69. Wang GL, Semenza GL.Characterization of hypoxia-inducible factor 1 and regulation of DNA binding activity by hypoxia. J Biol Chem. 1993; 268(29):21513-8.
70. Wang GL, Jiang BH, Rue EA, Semenza GL.Hypoxia-inducible factor 1 is a basic-helix-loop-helix-PAS heterodimer regulated by cellular O2 tension. Proc Natl Acad Sci U S A. 1995; 92(12):5510-4.
71. Wang GL, Semenza GL. Purification and characterization of hypoxia-inducible factor 1.J Biol Chem. 1995; 270(3):1230-7.
72. Webster KA, Discher DJ, Hernandez OM, Yamashita K, Bishopric NH.A glycolytic pathway to apoptosis of hypoxic cardiac myocytes. Molecular pathways of increased acid production. Adv Exp Med Biol. 2000; 475:161-75.
73. Wenger RH, Marti HH, Schuerer-Maly CC, Kvietikova I, Bauer C, Gassmann M, Maly FE. Hypoxic induction of gene expression in chronic granulomatous disease-derived B-cell lines: oxygen sensing is independent of the cytochrome b558-containing nicotinamide adenine dinucleotide phosphate oxidase. Blood 1996; 87(2):756-61.
74. Wenger RH. Mammalian oxygen sensing, signalling and gene regulation. J Exp Biol. 2000; 203 Pt 8:1253-63.

75. Wiener CM, Booth G, Semenza GL In vivo expression of mRNAs encoding hypoxia-inducible factor -1. Biochem Biophys Res Commun. 1996; 225(2):485-8.
76. Youngson C, Nurse C, Yeger H, Cutz E. Oxygen sensing in airway chemoreceptors. Nature. 1993; 365(6442):153-5.
77. Youngson C, Nurse C, Yeger H, Curnutte JT, Vollmer C, Wong V, Cutz E. Immunocytochemical localization on O2-sensing protein (NADPH oxidase) in chemoreceptor cells. Microsc Res Tech 1997; 37(1):101-6
78. Yu AY, Frid MG, Shimoda LA, Wiener CM, Stenmark K, Semenza GL. Temporal, spatial, and oxygen-regulated expression of hypoxia-inducible factor-1 in the lung. Am J Physiol. 1998; 275(4):L818-26.
79. Zhu H, Qiu H, Yoon HW, Huang S, Bunn HF. Identification of a cytochrome b-type NAD(P)H oxidoreductase ubiquitously expressed in human cells. Proc Natl Acad Sci U S A. 1999; 96(26):14742-7.
80. Zhu H, Bunn HF.Signal transduction. How do cells sense oxygen? Science. 2001; 292(5516):449-51.

CHAPTER 4

THE BRAIN IN CRISIS

1. INTRODUCTION

The brain has almost no ability to continue normal operations when its energy supply is compromised. In humans for example, tightening a neck cuff to a pressure of 600 mm Hg produces unconsciousness within 6 to 7 s, and slow wave EEG activity takes over in less than 10 s.[101] Neuronal activity is also depressed during incomplete ischemia when the cerebral blood flow falls to less than 50% of control values.[116] If anoxia/ischemia is extended for much longer irreversible damage results. However, since the brain can fully recover from the first few minutes of anoxia, the early functional failures are compatible with survival, and indeed some may be part of a defense strategy.

With a few exceptions the general pattern of degenerative changes seen during anoxia is probably common to the brains of all vertebrates, not only mammals, although the time scale may vary between species due to differences in body temperature and metabolic rate. There are also some regional differences. In the mammalian brain for instance, the hippocampus, the corpus striatum and the Purkinje cells of the cerebellum appear to be among the structures that are most vulnerable to anoxic and ischemic damage, and within the hippocampus the CA1 neurons are more sensitive than, for example, those of the CA3 subregion.[96]

This chapter will look at the mechanisms that protect the brain during the initial hypoxic period and discuss the causes of the destructive changes that follow. Fig. 4.1 summarizes the sequence of interconnected events that will be discussed, a sequence that finally leads to neuronal death.

The Brain Without Oxygen. Third Edition. Peter L. Lutz, Göran E. Nilsson and Howard M. Prentice.

2. ENERGY FAILURE

Under normoxic conditions more than 95% of brain ATP is produced aerobically, whereby each mole of glucose is oxidized by 6 moles of O_2 to yield, theoretically, 36 moles of ATP. Empirical measurements of the ATP yield from glucose suggest that a production of 25 moles of ATP from each mole of glucose is closer to the truth.[44]

THE ANOXIC BRAIN CATASTROPHE

1) Halt in oxidative ATP production ⇒ The ATP level falls

2) The ATP utilizing Na^+-K^+ pump slows down

3) Net outward leakage of K^+ ⇒ Progressive depolarization

4) General depolarization. Rapid outflow of K^+ and inflow of Ca^{2+} and Na^+

5) Amino acid and Ca^{2+} transporters driven by the Na^+ gradient slow down or stop

6) Release of excitotoxins such as glutamate, aspartate and dopamine

7) The excitotoxins activate receptors that cause additional inflow of Ca^{2+} and Na^+

8) Neuronal death caused by:

A) Activation of lytic processes (proteases, lipases and endonucleases) by Ca^{2+}

B) Cell swelling and lysis due to water accompanying the inflow of ions ⇒ Increased cerebral pressure ⇒ Permanent ischemia

C) Elevated levels of free fatty acids and other lipids cause membrane damage

D) Production of cytotoxic free radicals and aldehydes, especially upon reperfusion

Fig. 4.1 An overview of the sequence of interconnected events that finally leads to neuronal death in the vertebrate brain.

By contrast, anaerobic glycolysis – the breakdown of glucose to pyruvate or lactate – only yields 2 moles of ATP per mole of glucose. Under normal (fully aerobic) conditions, lactate production accounts for only a negligible (<4%) amount of the glucose metabolized,[22,75] thus most glucose consumed is fully oxidized to H_2O and CO_2. However, during anoxia, glycolysis is the only pathway by which ATP can be produced. Thus, to fully compensate for the loss of oxidative ATP p roduction, the anoxic brain would have to increase its glycolytic rate more than 10 times. Very few brains, if any, can accomplish this.

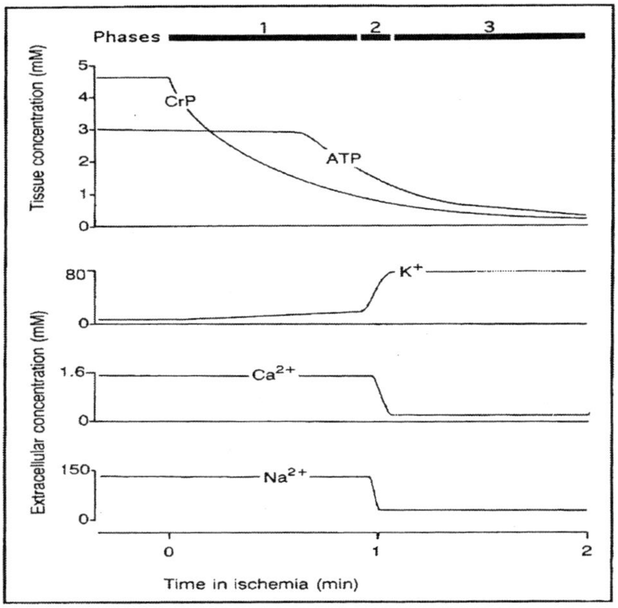

Fig. 4.2 Changes in ATP, creatine phosphate (CrP) and extracellular ion concentrations during ischemia in the rat brain. The changes in the extracellular levels of the ions are produced by their movement over the cell membrane and reciprocal changes in concentration occur simultaneously in the cytosolic compartment. The changes in ion distribution occur in three distinct phases[37]*. Phase 1 is characterized by a slow increase in extracellular* K^+*. In phase 2, which takes place after about 60 s of ischemia in the rat brain, rapid changes in ion distribution are seen. Phase 3 denotes the subsequent steady ion levels following phase 2. Compiled from Hansen*[37] *and Siesjo*[115].

Therefore, the paramount effect of anoxia in the brain, after it consumes its energy reserves, is energy failure. The brain's own reserves of oxygen and energy are trivial in comparison to its oxygen consumption. For example, compared with other organs, the brain has a low blood volume, only about 1% of it's wet weight, which under normoxic conditions consumption. For example, compared with other organs, the brain has a low blood volume, only about 1% of it's wet weight, which under normoxic conditions would hold about 0.09 μmol O_2/g brain tissue. On complete ischemia such a supply will only support normal oxygen consumption of about 1.6-5.0 μmol/g·min[114] for a few seconds. The reserves of energy-rich phosphate bonds (~P) are also very limited. For example, in the rat brain the average ATP concentration is about 3.0 μmol/g, which provides 6.0 μmol/g of ~P. The creatine phosphate concentration is 5.0 μmol/g, yielding 5.0 μmol/g of ~P. The total ~P concentration in the rat brain is therefore 11 μmol/g. At a normoxic consumption rate of about 50 μmol ~P / g·s, this store would only last about 20 s. The glycogen content of the mammalian brain is generally about 2-4 μmol/g which would support an aerobic glucose consumption rate of about 0.8 μmol/g·min for 2.5-5 min.[114] However, during anoxia the low efficiency of anaerobic glycolysis would mean that this store would only suffice for 9-18 s if normoxic ATP consumption rates were maintained. In anoxia or ischemia, the rapid fall in ATP production, without a comparable reduction in ATP utilization quickly leads to energy imbalance, a highly undesirable condition since most of the energy of the brain cells is spent on processes that are essential for neuronal functioning. In the mammalian brain the ATP concentration can be maintained for about 40 s of ischemia before falling precipitously, and complete depletion of ATP occurs within 1-2 min (Fig. 4.2).

In the most general way the consequences of energy deficiency in any biological system are related to the second law of thermodynamics, which states that energy is needed to maintain order in a system, i.e. to counteract increasing entropy. In the anoxic brain, this means that energy failure will threaten the structural integrity of the cell. Macromolecules are constantly being degraded in biological systems and energy is needed for their replacement. As we will see, the breakdown of macromolecules is actually accelerated during anoxia because many of the degrading enzymes are activated by Ca^{2+}, which rises steeply in intracellular concentration during anoxia.

Energy failure also means the loss of ion gradients. When the ATP level starts to fall, the ATP driven Na^+/K^+ pump can no longer counteract the downhill flow of ions through ion channels in the neuronal membrane and changes in the ion distribution are rapidly seen. This is one of the most intensively studied effects of anoxia/ischemia in the brain, and it is clearly a key event, since it happens very quickly and apparently precedes, and mediates, major perturbations of the macromolecular integrity of the cells. Indeed, most of the structural damage that is seen in the brain after an anoxic/ischemic insult is probably initialized by the loss of ion gradients. It should be pointed out that the Na^+/K^+ pump is not only essential for building up the membrane potential. The Na^+ gradient created by this pump is also the driving force for crucial processes like Ca^{2+} extrusion (via the Na^+/Ca^{2+} exchanger), uptake of organic molecules such as amino acid neurotransmitters, and regulation of cell volume.[73]

The consequential major factors that appear to be involved in ischemic/anoxic brain damage are: a net outflow of excitatory neurotransmitters, a massive rise in intracellular Ca^{2+} due to a failure of Ca^{2+} regulating mechanisms, acidosis related to the increased lactate from anaerobic glycolysis, the production of nitric oxide and other free radicals, lipolysis, and cell swelling.

Based primarily on the major ion movements displayed by the anoxic brain[38], the brains responses to anoxia can be divided into 3 distinct phases (Fig. 4.2). Initially (phase 1) there is a rapid decrease in electrical activity and a suppression of neurotransmission. During this phase there is a slow increase in extracellular K^+. After only a few minutes phase 2 abruptly follows, characterized by rapid depolarization, a loss of ion gradients, and a massive outflow of various neurotransmitters, of which some are highly excitotoxic at these elevated extracellular concentrations. In phase 3, the membrane is fully depolarized and if this last phase is prolonged for more than a few minutes, a sequela of irreversible damage ensues.

Clearly, an understanding of these processes is a key to providing effective protective or remedial intervention strategies. What do we know of the mechanisms that induce and maintain the first, probably defensive phase, what are the threshold conditions that trigger the second "pathological" phase, and what are the proximal causes of the destructive consequences?

3. PHASE 1: METABOLIC DEPRESSION, A MECHANISM FOR PROTECTION

The rapid interruption in synaptic transmission that occurs early during hypoxia could be due to several causes, including the failure of presynaptic impulses to release neurotransmitters, changes in ion conductance, altered properties of receptors, and a reduction in postsynaptic excitability. The suppression of synaptic transmission is probably due to a combination of several effects. It has been suggested, for example, that the initial failure of neurotransmitter release during hypoxia is due to a blockade of voltage-gated Ca^{2+} channels, and with the normal uptake of Ca^{2+} blocked neurotransmitters can no longer be released.[123] The inactivation of the voltage dependent Ca^{2+} currents may be caused by an increase in intracellular Ca^{2+} since Ca^{2+} is released from intracellular stores during hypoxia.[63] Another factor is the several minutes of hyperpolarization that many neuronal tissues exhibit during anoxia. Hyperpolarization would cause a depression of neuronal excitability by increasing the excitatory input needed to reach the action potential threshold. In hippocampal pyramidal neurons the anoxic hyperpolarization, which is accompanied by a 50% fall in membrane resistance, may be sufficient by itself to account for the strong depression of excitability in this tissue.[63] The combination of hyperpolarization and a falling membrane resistance is thought to be due to an activation of K^+ channels which produce a membrane potential shift towards the K^+ Nernst equilibrium potential. At this time inward currents of Ca^{2+} and Na^+ are also depressed.

There is evidence that during this period there is an influx of Na^+ via the voltage-dependent Na^+ channels.[132] It has been suggested that this movement of Na^+ is accompanied by Cl^- and water with consequent cell swelling and a corresponding reduction in the extracellular space.[73] A reduction in the extracellular space by about 20% during phase 1 was observed by Hansen and Olsen.[39]

The decrease in neuronal activity provides energy savings, which may be considerable. As discussed in the previous chapter there is evidence that the electric activity of the brain is responsible for 50% or more of the brain energy consumption. The hyperpolarization stage can therefore be seen as a protective response since it allows the brain to "hold on" and maintain ionic gradients and membrane potential in the face of a decreasing energy supply, albeit for only one or a few minutes.

However, this strategy is essentially short term since it plants the seeds of its own destruction. The increased K^+ permeability produces a net outward leakage of K^+ from the neurons and a slow but steady increase in extracellular K^+ levels. As the extracellular K^+ rises the membrane potential becomes more depolarized, and soon a destabilizing threshold is reached.

As might be expected considerable interest has been directed towards the mechanisms that are responsible for this increase in K^+ conductance. At least two types of K^+ channels appear to be involved, one activated by an increase in intracellular Ca^{2+} the other activated by a decrease in ATP (K_{ATP} channels).[63]

The activation of Ca^{2+} sensitive K^+ channels is thought to be responsible for the early suppression of the EEG which occurs within 15 s of ischemia, the initial rise in cytosolic Ca^{2+} being due to a release from intracellular stores since extracellular Ca^{2+} does not appear to decrease during this period.[121]

K_{ATP} channels may mediate hyperpolarization in at least certain brain regions, such as the hippocampus and brain stem,[33,51,85] although there is some controversy since at the time when synaptic transmission is abolished only marginal changes are seen in brain ATP concentrations.[36] However, K_{ATP} channels may also be activated by increasing ADP levels, which rise 2-fold and 4-fold in rat cortex after 15 s and 45 s of ischemia, respectively.[20] Moreover, there is evidence that more substantial reductions in ATP occur in the areas of the hippocampus where most of the excitatory synapses are concentrated (the "molecular layer").[74] In the hippocampus at least, K_{ATP} channels are mainly present in glutamate releasing axon terminals, and may serve to reduce the release of this excitatory neurotransmitter early on in anoxia.[62,67] Interestingly, both K_{ATP} channels and Ca^{2+} sensitive K^+ channels have been identified in the amphibian sciatic nerve where they may act synergistically to protectively decrease cell excitability during hypoxia.[53]

By using a specific blocker of K_{ATP} channels, glibenclamide, Jiang et al.[52] obtained evidence that in brain-stem hypoglossal neurons of adult rats about 50% of the increase in extracellular K^+ during phase 1 was caused by the opening of K_{ATP} channels. However, they found no signs of an opening of K_{ATP} channels in the brains of newborn rats and freshwater turtles during anoxia. Both freshwater turtles and newborn rats are much more anoxia tolerant than adult rats and the functional aspects of these findings will be further discussed in the next chapter.

4. PHASE 2: COMPLETE ENERGY FAILURE AND ANOXIC DEPOLARIZATION

4.1 Ion Movements

During phase 1, there is a small (18% elevation) increase in extracellular Ca^{2+} and a relatively slow, but greater increase (about 160%) in extracellular K^+ The increase in Ca^{2+} is thought to result mainly from a decrease in the extracellular space[132] the much greater increase in K^+ is due to a net outflow of K^+ from the neurons. When the concentration of K^+ reaches about 8-10 mM an abrupt transition takes place: within a few seconds, extracellular K^+ rises steeply to 50-70 mM while extracellular Na^+ and Ca^{2+} concentrations fall as these ions flood into the neurons (Fig. 4.2). It has been suggested that Na^+ influx during phase 1 is a major factor leading to cell depolarization.[132] Application of tetrodotoxin (TTX) a specific antagonist for the voltage-dependent Na^+ channel significantly increases the length of phase 1, (from 207 s to 342 s at 32 °C) in the ischemic isolated perfused rat brain, and delays the corresponding ion changes.

Energy supply is a key to these changes, as seen in the accelerated rate of increase in extracellular K^+ during anoxia following blocking glycolysis with iodoacetate,[131] and the delay in the time to reach the phase 2 threshold by infusing a hyperglycemic saline.[10,37] In mammals the ischemic threshold for this loss of ionic homeostasis occurs at cerebral blood flow rate below 0.10-0.12 ml/g·min.[114]

This pattern of an initial slow increase in extracellular K^+ to a threshold that triggers a rapid rise in extracellular K^+ appears to be a general response of vertebrate brains to anoxia or ischemia, although the time to reach the threshold may vary between species according to their body temperature. Thus, in rats the threshold is reached within 1-2 min of ischemia or cardiac arrest[34,114] and within 2-3 min of anoxia (N_2 respiration)[115], while in rainbow trout held at 10 °C, an abrupt rise in brain extracellular K^+ is seen after about 30 min of anoxia (see Fig. 6.3B in Chapter 6). However, if temperature differences are taken into account the time taken for the anoxic rainbow trout brain to reach phase 2 is similar to that of the mammal, i.e. 30 min at 10 °C is approximately analogous to 4-5 min at 37 °C.[88]

The rapid increase in extracellular K^+ corresponds to a general depolarization of the brain, often denoted anoxic depolarization. As illustrated by Fig. 4.2, apart from the drastic increase in extracellular K^+, there are simultaneous floods of extracellular Na^+ and Ca^{2+} into the neurons that are probably more important than the movement of K^+ for the development of anoxic cell death.

The exact nature of the mechanisms responsible for this sudden increase in membrane permeability to K^+, Na^+ and Ca^{2+} during phase 2 is not well understood though several mechanisms have been suggested. One of the most straightforward explanations is that the membrane depolarization triggers the opening of voltage-gated K^+, Na^+ and Ca^{2+} channels – the same channels that are responsible for the generation and termination of action potentials and initiation of neurotransmitter release. However, experiments where one or more of these channels have been blocked indicate that the opening of voltage-gated ion channels is not the only mechanism by which the neuronal membrane opens up to ions during phase 2.[16,52] Other possible ways of entry for these ions are through ligand gated ion channels, notably glutamate gated non-selective cation channels such as the NMDA receptor. There are results indicating that the NMDA receptor can open its ion channel in response to cellular energy failure due to a loss of one or more intracellular mechanisms that regulate the state of the channel.[125] Thus channel opening may occur in the absence of any net increase of the endogenous levels of NMDA receptor ligands – glutamate and aspartate. Other possible mechanisms mediating anoxic depolarization include a reversal of ion exchangers such as the Na^+/Ca^{2+} exchanger, which could facilitate Ca^{2+} moving into the cells as a result of the inversion of the normal Na^+ gradient.[73]

Finally, Somjen and co-workers[52,122] have suggested the existence of stretch activated ion channels that open in response to cell swelling that occurs during anoxia. Recently, such a stretch activated channel was proven to exist in cultured rat neurons Kim et al.[58], and stretch activated channels have previously been found in other tissue s.[98] It is possible that many of the known channel types (e.g. receptor or voltage gated channels) could open indiscriminately in response to stretching during cell swelling.[122] The channel characterized by Kim et al.[58] was a K^+ channel that opened in response to fatty acids (such as arachidonic acid) as well as membrane stretch, and the authors suggested that it could be involved in both cell volume regulation and the ischemia-induced elevation of neuronal K^+ loss.

4.2 *Neurotransmitter Release*

Another drastic event that characterizes the anoxic vertebrate brain is a massive release of neurotransmitters into the extracellular space where they interact with their receptors and activate receptor gated ion channels.

A technique that has proven to be particularly useful for investigating these processes is microdialysis (Fig. 4.3), which allows continuous monitoring of extracellular neurotransmitter levels in vivo.[129] Similar measurements can also be made using "cortical cups" to sample extracellular fluid leaking out from surface of the mammalian cerebral cortex.[95] Most studies have focused on amino acids. During ischemia or anoxia, a release of glutamate (Fig. 4.4), aspartate, GABA, glycine and taurine has been unequivocally demonstrated to occur in the several areas of the adult and neonatal mammalian brain, including cerebral cortex, hippocampus, thalamus and corpus striatum.[28,34,93,95] A rapid release of amino acids during phase 2 of anoxia appears to be a general phenomenon among vertebrates since the same amino acids are released in the thalamus of the anoxic rainbow trout brain.[47]

One question that has been difficult to resolve is how the time course of neurotransmitter release relates to the major ion movements during anoxia.[84] In other words, is the release of neurotransmitters a consequence of the anoxic depolarization, or is it one of the mechanisms that cause anoxic depolarization? Unfortunately, because the time resolution of microdialysis and related techniques is generally 5-20 min or more, this question has been difficult to answer unequivocally with regard to amino acid neurotransmitters. In addition, the microdialysis probe is relatively large (about 250 µm in diameter or more) and cannot reach into the synaptic cleft (only a few nm wide). Thus, neurotransmitters have to diffuse a considerable distance from their release sites to reach the probe, which will cause a time lag between neurotransmitter release and its detection by the microdialysis. In a study where the microdialysis sampling periods were kept exceptionally short (1 min), the results suggested that the release of glutamate starts during the anoxic depolarization (phase 2).[55]

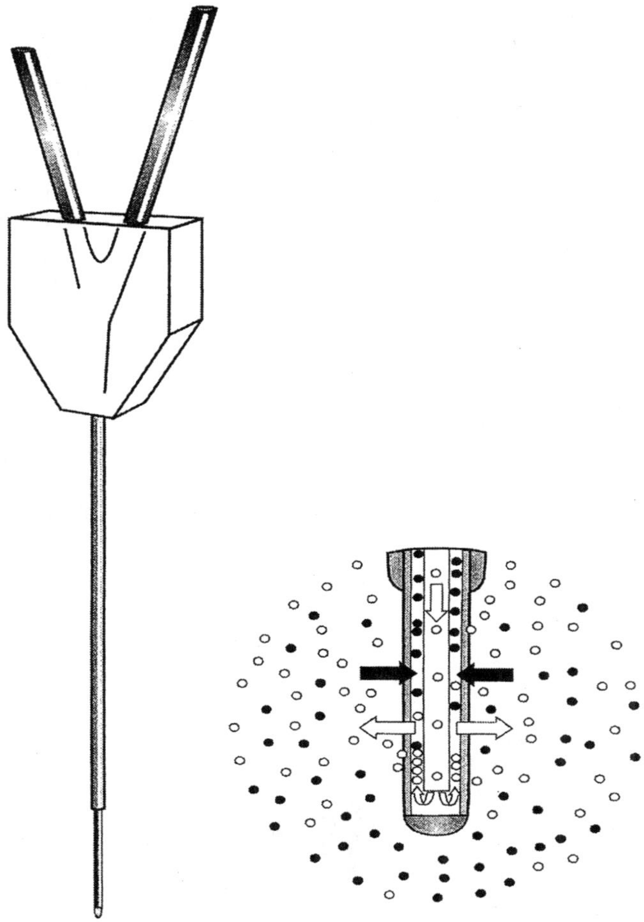

Fig. 4.3 A microdialysis probe used to monitor extracellular neurotransmitter levels in vivo. The diameter of the probe is 0.25 mm. When a saline solution is pumped through the inner tube of the probe, small molecules will diffuse into this solution over the dialysis membrane (lower right). The saline that is brought back by the outer tube is sampled regularly and analyzed with regard to its content of neurotransmitters. Courtesy of CMA/Microdialysis AB (illustrator: Lena Kehr).

The monoamine neurotransmitters, dopamine, norepinephrine and serotonin, are also released into the extracellular space of the mammalian brain during anoxia[29] or even mild hypoxia (Fig. 4.5).[46] However, changes in monoamine concentrations can be followed by in vivo voltammetry, a technique that allows a much faster time resolution than microdialysis. Using this technique, Moghaddam et al.[84] were able to show that the dopamine release in corpus striatum during anoxia is already underway in phase 1, at about the same time as extracellular K^+ starts to rise, and the extracellular dopamine level is close to its peak value when the anoxic depolarization (phase 2) occurs (Fig. 4.6). It could be that some of the amino acid neurotransmitters are also released at the same time as dopamine, before anoxic depolarization. However, not all amino acid neurotransmitters are released at the same time. For example, some studies have indicated that the extracellular concentration of GABA starts to increase a few min before that of glutamate in the anoxic or ischemic mammalian brain.[60,93,95] Comparative experiments on an ectothermic vertebrate, the rainbow trout, have offered new insight into the timing of neurotransmitter release.

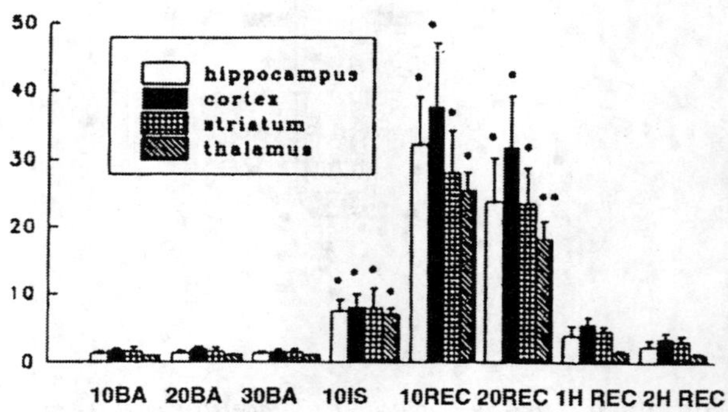

Fig. 4.4 Time course of changes in the microdialysis perfusate levels (µM) of glutamate sampled from various brain regions of the rat subjected to 20 min ischemia. BA, baseline; IS, ischemia; REC, recovery (from Globus[31]).

The reason for this is that the low body temperature of the ectothermic trout greatly extends the time frame during which these events occur. This microdialysis based study[47] showed that in the thalamus of rainbow trout held at 15 °C, anoxia exposure results in the release of several amino acid neurotransmitters, including glutamate, aspartate, GABA, glycine and taurine (Fig 4.7).

In contrast to the mammal, GABA release in the trout appeared to be delayed compared to glutamate. Interestingly, the loss of ion-homeostasis (measured by K^+ selective microelectrodes) started after 23 min, long before the earliest release of any of the amino acids (starting after 45 min). The amino acid release had no apparent effect on the rate of increase in extracellular K^+. Thus, in the rainbow trout, if these events are interrelated, the loss of ion-homeostasis is likely to trigger the amino acid release, but not vice versa.[47]

So what mechanisms are responsible for release of amino acids during anoxia? Normally, the route for neurotransmitter release is the fusion of transmitter-containing synaptic vesicles with the presynaptic neuronal membrane, initialized by Ca^{2+} flowing in through voltage-gated Ca^{2+} channels in response to membrane depolarization. However, this route for neurotransmitter release is probably more or less closed during anoxia, because this mechanism is inhibited by a fall in cellular energy charge. The reason for this is that the fusion of the vesicles with the neuronal membrane demands ATP.

An alternate route for neurotransmitter release during anoxia and ischemia has been suggested and has also received experimental back up. This route involves a stop or reversed operation of amino acid reuptake. Several amino acids, including glutamate, aspartate, GABA, glycine and taurine, are taken up by neural cells through electrogenic high-affinity Na^+/amino acid co-transporters[79,126] These transporters are responsible for the reuptake of the amino acid neurotransmitters that have been released from synapses, and the capacity of these transporters to keep the extracellular levels of amino acid neurotransmitters low is of course a prerequisite for synaptic transmission. These co-transporters, which are normally driven by the Na^+ gradient and the membrane potential, transport the negatively charged amino acid molecules together with Na^+ ions into the cell. Because these uptake routes are dependent on the Na^+ gradient over the cell membrane, the loss of the Na^+ gradient after depolarization will make the Na^+-amino acid co-transporters unable to remove amino acids from the extracellular space.

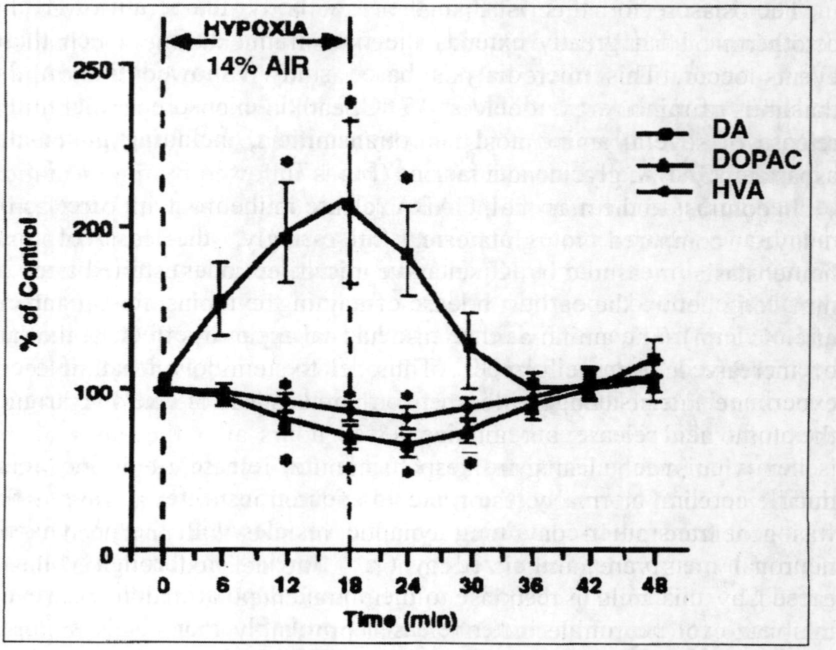

Fig. 4.5 Effect of mild hypoxia (14% air) on the extracellular concentrations of dopamine and metabolites (DOPAC, HVA) in the striatum of piglets (from Huang et al.)[46]

The glutamate carrier normally transports one glutamate anion and Na^+ into the cell while one K^+ and one OH^- or HCO_3^- are transported out of the cell.[125] The direction of the glutamate transport is determined by concentration gradient of other the ions and the ion changes in anoxia/ischemia actually cause this transporter to reverse and release glutamate from the cell. Stoichiometric calculations suggest that following anoxic depolarization the glutamate uptake carrier will pump glutamate out of the cell until a new equilibrium is reached at a neurotoxic level of greater than 100 μm.[125]

Nevertheless, there appears to be an early release of glutamate that occurs before depolarization and is not dependent upon extracellular Ca^{2+} entry. This glutamate has been suggested to come from the metabolic pool[49], but alternatively it may be stimulated by an increase in intracellular Ca^{2+} level due to a release from intracellular Ca^{2+} stores.[56] Thus, the normal vesicular (Ca^{2+}- and ATP-dependent) glutamate release could in some situations make a significant contribution to the increase in extracellular glutamate levels seen early during ischemia or anoxia, before it is inhibited by the fall in cellular energy charge.[55,106] Moreover, after ischemia, when neuronal ATP levels starts to rise, glutamate is released by Ca^{2+}-dependent exocytosis which in turn activates an excessive influx of Ca^{2+} largely through potentiated NMDA-receptor-channels.[125]

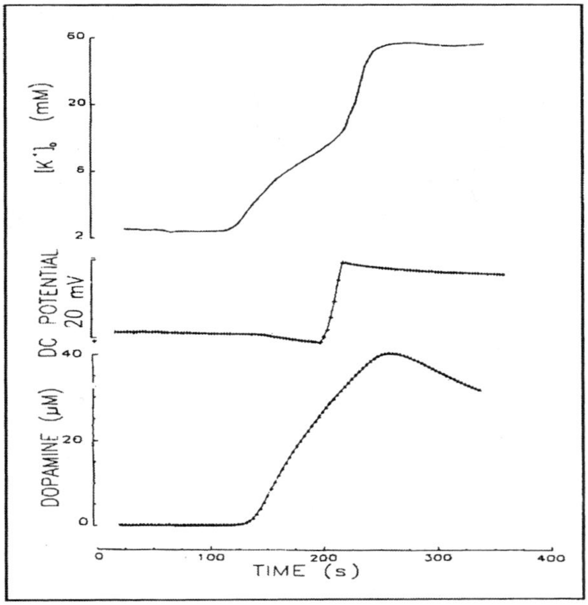

Fig. 4.6 Simultaneous recordings of extracellular K^+, brain DC potential and extracellular dopamine in rat striatum during ischemia. Dopamine was measured using in vivo voltammetry. Note that dopamine is released before the phase 2 K^+ outflow starts and before the brain depolarizes (as shown by the DC potential). From Moghaddam[84], with permission National Research Council of Canada.

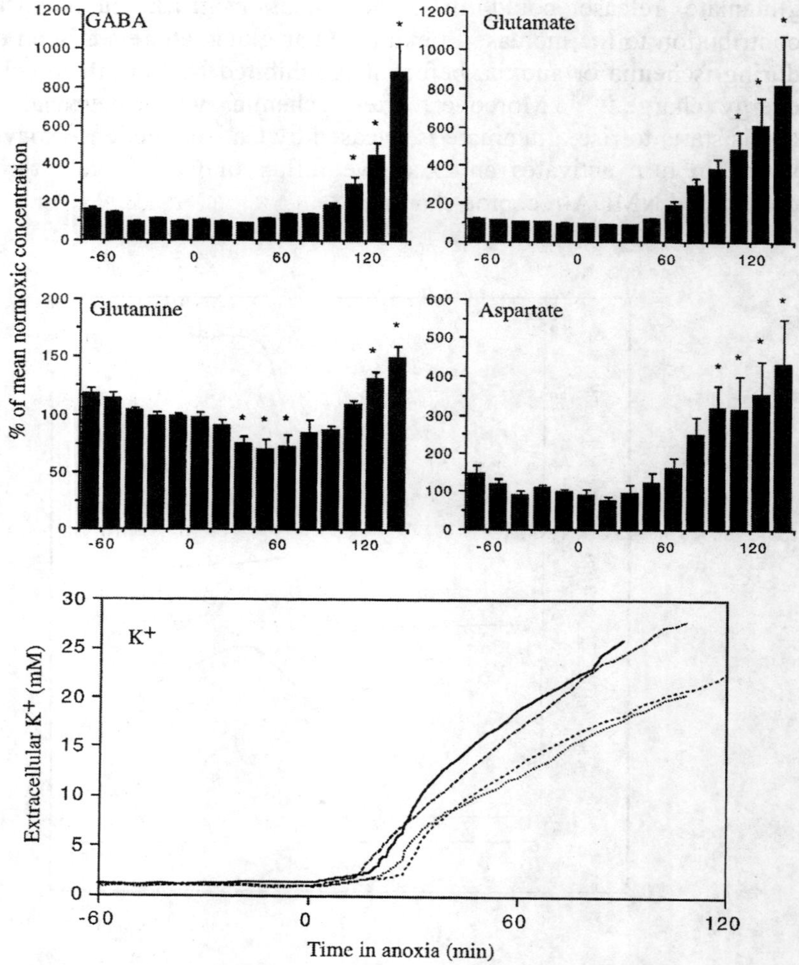

Fig. 4.7 The anoxic rainbow trout brain: time course of changes in extracellular K^+, GABA, glutamate, glutamine and aspartate sampled from thalamus before and during anoxia. K^+ was measured with an extracellular ion-selective electrode, while the amino acids were sampled with microdialysis. From Hylland et al.[47]

5. PHASE 3: NEURONAL DESTRUCTION

5.1 Neurotransmitters Become Excitotoxins

The explosive rise in the extracellular levels of excitatory neurotransmitters, in particular glutamate, aspartate and dopamine, is though to be a major cause of anoxic/ischemic neuronal destruction since these neurotransmitters become excitotoxins at high extracellular concentrations.

Glutamate plays a particularly important role in producing brain damage through its actions as an agonist of both AMPA and NMDA receptors, and it has been suggested that activation of AMPA receptors by glutamate during anoxia sets the stage for the activation of NMDA receptors.[73] When glutamate activates the AMPA receptor, an ion channel with conductance for the monovalent cations Na^+ and K^+ is opened, thereby depolarizing the neuron. The depolarization in turn allows glutamate to activate the NMDA receptor (NMDA channel) by removing a voltage dependent block on this receptor. Thus, an initial depolarization of the neuronal membrane caused by an activation of AMPA gated ion channels (or other channels with Na^+ conductance) leads the way for glutamate to activate NMDA receptors, which, apart from allowing the entry of Na^+ and K^+ also have a high conductance for Ca^{2+}, causing Ca^{2+} ions to flood into the cell, with devastating consequences.

Indeed, the toxicity of glutamate during anoxia appears to be tightly connected to the ability of glutamate to stimulate Ca^{2+} inflow to the neurons through NMDA channels. For example, in cultured rat hippocampal neurons, exposure to a high glutamate concentration for 30 min is only deleterious if Ca^{2+} is present in the culture medium.[103] However, it should be mentioned that glutamate exposure in the absence of Ca^{2+} still induces cell swelling – a potentially highly deleterious process (see below). A "domino effect" threatens in this process, the failed neurons releasing neurotoxic amounts of excitatory neurotransmitters which stimulate neighboring vulnerable cells to release their neurotransmitters, with further fatal consequences and a resultant spreading wave of death. Possibly due to impaired uptake mechanisms, glutamate, aspartate and dopamine may remain at high concentrations, or rise even further, for several hours after circulation has been reestablished following ischemia.[2,7,93] This may be an important cause of the delayed anoxic/ischemic brain damage that occurs after reperfusion of the tissue with oxygenated blood – a somewhat paradoxical situation called reperfusion damage.

However, delayed neuronal death has repeatedly been observed long after excitatory neurotransmitter levels have been restored. Still, excitatory neurotransmitters may be responsible: A recently suggested (and still largely hypothetical) glutamatergic mechanism that may contribute to delayed anoxic brain damage is related to long-term potentiation (LTP), a phenomenon that appears to be strongly dependent on the activation of NMDA receptors.[90] LTP is the long-lasting enhancement of synaptic transmission in response high-frequency stimulation of synapses. Therefore, it is not difficult to imagine that the complete depolarization and massive glutamate release occurring during anoxia/ischemia may induce LPT and potentiate a massive Ca^{2+} influx that could have severe consequences for neuronal survival long after an ischemic/anoxic insult.

Excitatory neurotransmitters may be a factor in the regional differences in anoxia vulnerability seen in the brain.[30] For example, the greater vulnerability of corpus striatum to anoxia/ischemia damage has been related to the comparatively slow postischemic decline (recovery) in extracellular levels of glycine (an allosteric activator of the NMDA receptor) in this brain region.[31] Clearly a knowledge of the mechanisms that account for the differences in anoxia tolerance of different (sub)regions of the vertebrate brain may help to clarify the primary causes of hypoxic damage.

With all the interest directed against the NMDA receptor and its excitotoxic role in anoxic/ischemic brain damage, there have of course been numerous studies aimed at evaluating the usefulness of NMDA receptor antagonists as protective agents against ischemic neuronal damage in animal models. One puzzling aspect of these studies is that while NMDA receptor antagonists have repeatedly proven to substantially attenuate the injury caused by focal ischemia in mammals[102], the same drugs fail to limit global ischemic brain damage. Instead, during global ischemia, AMPA receptor antagonists have been reported to provide protection.[86] The reason for this remains unclear and shows that much is still to be learned about the role of glutamate in anoxic/ischemic brain damage.

Not only amino acid neurotransmitters are released during anoxia/ischemia in the mammalian brain. Extracellular levels of ACh and monoamines (serotonin, norepinephrine and dopamine) also increase excessively during ischemia.[28,29,30,64,90] This occurs in spite of a decrease in the tissue levels of these neurotransmitters. As mentioned, the synthesis of monoamines have an absolute demand for molecular oxygen, and the synthesis of ACh is also reduced during anoxia, probably because the stop in the oxidative degradation of glucose results in a shortage of acetyl-coenzymeA needed for ACh synthesis.[90] Like glutamate, the release of monoamines has been implicated in neuronal damage during ischemia, because all of them function as excitatory neurotransmitters in certain neuronal systems in brain.[29,40,90] For example, in some brain areas, particularly in the striatum, dopamine is an excitatory neurotransmitter and one effect of dopamine is to inhibit Na^+/K^+ pump activity, thereby causing a rise in intracellular Na^+.[5]

Particular attention has been directed to the possibility that the increase in extracellular dopamine seen in ischemic/anoxic neuronal tissues may also be an important contributing factor to neuronal damage by modulating the release of excitatory amino acids, generating free radicals and uncoupling CBF and glucose metabolism.[28] The increase in brain extracellular dopamine can be massive, as much as 250 fold following 20 min ischemia in the rat striatum.[107] However, in contrast to glutamate and many other amino acid neurotransmitters, there sometimes appears to be no critical level or O_2 reserve which holds back a dopamine release. In newborn piglets even a mild reduction in cortical oxygen pressure (from a normoxic PO_2 level of 31-35 torr to a hypoxic 24 torr) produced an 80% increase in dopamine and the greater the hypoxia the greater the increase in dopamine.[46] The level of extracellular dopamine is thought to be determined by the balance between the rate of active dopamine uptake, the rate of release and the rate of washout from the brain. The main route for removing released dopamine is cell inward transport and there appear to be two uptake processes – one high affinity and one low affinity. The high affinity uptake is inhibited by extracellular K^+ and is therefore likely to be suppressed during anoxia when extracellular K^+ is elevated.[6]

In severe hypoxia there is a massive rise in dopamine resulting from a large increase in dopamine efflux following depolarization, and a concomitant inhibition of reuptake. In mild hypoxia the main reason for the increase in extracellular dopamine is thought to be the inhibition of uptake.[46] Extracellular concentrations of the dopamine metabolites DOPAC and HVA are also hypoxia sensitive, declining markedly even under mild hypoxia (Fig 4.5). This fall possibly reflects a decrease in the activity of the oxygen dependent enzyme MAO.

However, in contrast to glutamate, the monoamines function as inhibitory neurotransmitter in certain neuronal systems of the brain, where they could have a protective effect during anoxia. Indeed, in some ischemic models, neuroprotection has been seen with drugs stimulating dopaminergic, serotonergic or cholinergic neurotransmission.[40,90]

5.2 *Intracellular Calcium*

The sudden increase in extracellular K^+ is associated with a rapid fall in extracellular Ca^{2+} (Fig. 4.2) as Ca^{2+} floods into the cell. This uncontrolled increase in intracellular Ca^{2+} appears to play a critical role in the destructive events that follow. In the normal life of a neuron, changes in intracellular Ca^{2+} serve a multitude of functions, Ca^{2+} being a 2nd messenger and a regulator of several key metabolic pathways. As such it is normally tightly controlled, the intracellular concentration of free Ca^{2+} being maintained at about 10^{-7} M, i.e. 1/10,000 of the extracellular level.

As mentioned above, there are several mechanisms underlying the anoxic Ca^{2+} influx, including the opening of voltage gated Ca^{2+} channels, activation of agonist gated Ca^{2+} channels such as the NMDA glutamate receptor, and reversal of the Na^+/Ca^{2+} exchanger – which is normally the principal Ca^{2+} extrusion mechanism. Unfortunately, the inflow of Ca^{2+} can be self-amplifying. At the presynaptic terminal the increase in intracellular Ca^{2+} facilitates the release of the excitatory amino acids glutamate and aspartate from presynaptic terminals. At the depolarized postsynaptic membrane the released glutamate activates the NMDA receptor and further increases Ca^{2+} permeability. In addition, the fall in intracellular ATP causes a net release of bound Ca^{2+} from the endoplasmic reticulum and mitochondria, since both these organelles demand ATP (and O_2) for their Ca^{2+} sequestration.[10,73] Moreover, mitochondria are relatively sensitive to ischemic insults (see below) and by leaving the cell with an impaired capacity for ATP production and Ca^{2+} sequestration the damaged mitochondria may underlie some of the delayed neuronal death seen after an anoxic/ischemic episode.[3,42]

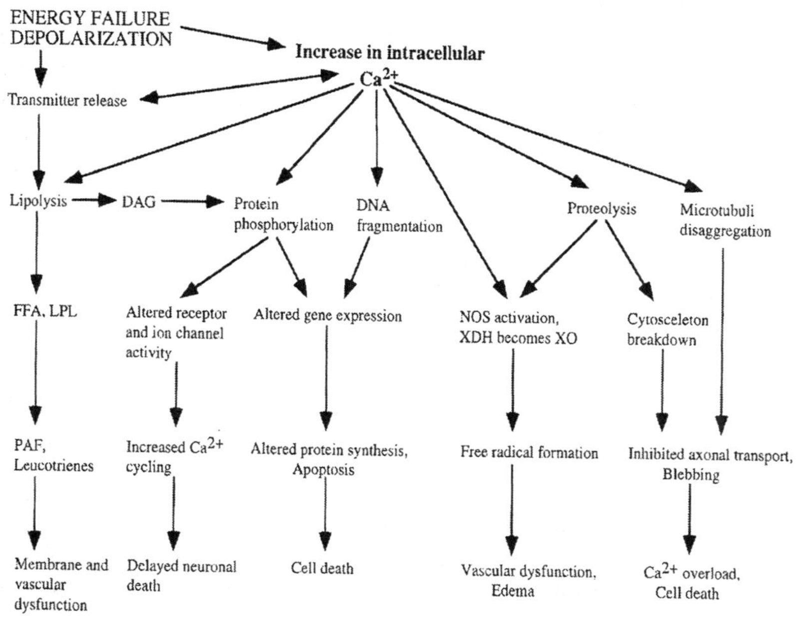

Fig. 4.8 Proven and possible mechanisms by which a rise in intracellular Ca^{2+} jeopardizes neuronal survival during and after anoxic/ischemic energy failure. Note the interrelationships indicated between several of the potentially deadly mechanisms discussed in this chapter, which include lipolysis, neurotransmitter release, apoptosis, free radical formation and NO production. (DAG = diacylglycerides; FFA = free fatty acids; LPL = lysophospholipids; NOS = nitric oxide synthase; PAF = platelet activating factor. XDH = xanthine dehydrogenase; XO = xanthine oxidase). Modified from Siesjö[119] *and Siesjö et al.*[116]

The uncontrolled and explosive rise in intracellular Ca^{2+} that follows upon anoxia or ischemia acts as a trigger for multiple dysfunctional effects (Fig. 4.8), including the activation of enzymes such as phospholipases and proteases, leading to a degradation of major cellular components, as well as kinases and phosphatases, leading to a dysfunction of receptor function, gene expression and protein synthesis. A consequence of the stimulated phospholipid hydrolysis is the formation of toxic metabolites and damaging free radicals.

Indeed, it appears that the elevation of intracellular Ca^{2+} initiates several processes that will continue, or actually speed up neuronal destruction even after the brain is reperfused, acting as further agents in reperfusion damage. It was relatively recently shown that blocking voltage-gated Ca^{2+} channels (of the N-type) during reperfusion with a synthetic conopeptide (SNX-111) significantly attenuates ischemic damage in the hippocampus of rats, even if the conopeptide is administered up to 24 h after the 15 min long ischemic insult.[130]

However, a rise in intracellular Ca^{2+} is probably not the whole story. Rothman and Olney[102] have pointed out that, although large increases in intracellular concentrations of Ca^{2+} are seen in neurons exposed to toxic levels of glutamate and NMDA, other manipulations that also produce high levels of intracellular Ca^{2+} may only result in minor or no cell injury.

6. NITRIC OXIDE PRODUCTION

There is accumulating evidence for a strong involvement of nitric oxide (NO) in the pathophysiology of cerebral ischemia. Experimental results on the involvement of NO in these events often appear to be conflicting, some data suggesting that NO production is primarily detrimental[11,89] while others indicate a protective role for NO in cerebral ischemia.[133,135,137] However, there has been success in reconciling these seemingly conflicting results.[17,136] The key to understanding the role of NO in brain is to recognize that there are several NO producing systems in brain, some being activated immediately upon hypoxia/ischemia and some responding in a delayed fashion. Moreover, some NO effects will be of benefit for hypoxic/ischemic brain survival while other are probably highly detrimental. The involvement of NO in brain hypoxia/ischemia is summarized in Fig. 4.9. We are purposely discussing hypoxia instead of anoxia here, since under conditions of complete anoxia, NO can not be produced due to the O_2 requirement of NO synthesis. However, during the periods of hypoxia before and after an anoxic event, NO is certainly involved.

Two main sources of NO can be identified: constitutive NOS

causing a rapid NO production in neurons and endothelial cells, and inducible NOS responsible for a delayed but more prominent NO production. The immediate rapid mechanisms responsible for NO production in the ischemic/hypoxic brain appear to be closely related to two of the main events that we have discussed so far in this chapter: the activation of the NMDA receptor and increased intracellular Ca^{2+} levels. In neurons, the inflow of Ca^{2+} caused by NMDA receptor activation will lead to increased NO production from Ca^{2+} dependent neuronal NOS (NOS-I). Activation of neuronal NOS can of course also occur in response to Ca^{2+} rises initiated by receptors for neurotransmitters other than glutamate or other mechanisms that promote increased intracellular Ca^{2+} levels during hypoxia/ischemia (such as an opening of voltage gated Ca^{2+} channels and a reversal of the Na^{+}/Ca^{2+} exchanger). Since endothelial NOS (NOS-III) is also activated by increased intracellular Ca^{2+} levels, this is another likely source of NO during brain energy deficiency. Experiments on focal ischemia have shown that there is a rapid rise in NO levels during the first minutes, and that these fall back to the baseline within an hour.[53,80]

A secondary but much more powerful production of NO can come from micro-glia cells and invading macrophages which both contain inducible NOS (NOS-II). This source of NO may become important several hours or days after an ischemic episode, when these cells start to occur at high numbers in the brain.[17] It may be noted that endothelial NOS has also been found to be slowly induced by focal ischemia in rats: the enzyme level (determined by monoclonal antibodies specific for NOS-III) was increased 5 fold after 24 h and remained at an elevated concentration for several days.[139] Subsequently, there is a rapid upregulation of mRNA for neuronal NOS-I after an ischemic lesion, an increase that can be detected as early as 15 min after MCA occlusion.[140]

In the normoxic or non-ischemic situation, several studies have shown that the neurotoxicity of NMDA receptor activation can be attenuated by NOS inhibitors both in vitro and in vivo.[17,32] In fact, NOS inhibitors can be as effective in protecting against NMDA neurotoxicity as MK-801 – a blocker of the NMDA receptor itself.[26] This clearly suggests that NO production in brain can be detrimental. However, in focal ischemia, the ability of NOS inhibitors to reduce brain damage is much less than that regularly seen with NMDA blocker.[17] This can probably been ascribed to the beneficial effect that NO may have on the outcome of cerebral ischemia. Thus, we have now to look more closely at the effects that NO may have in the hypoxic/ischemic brain.

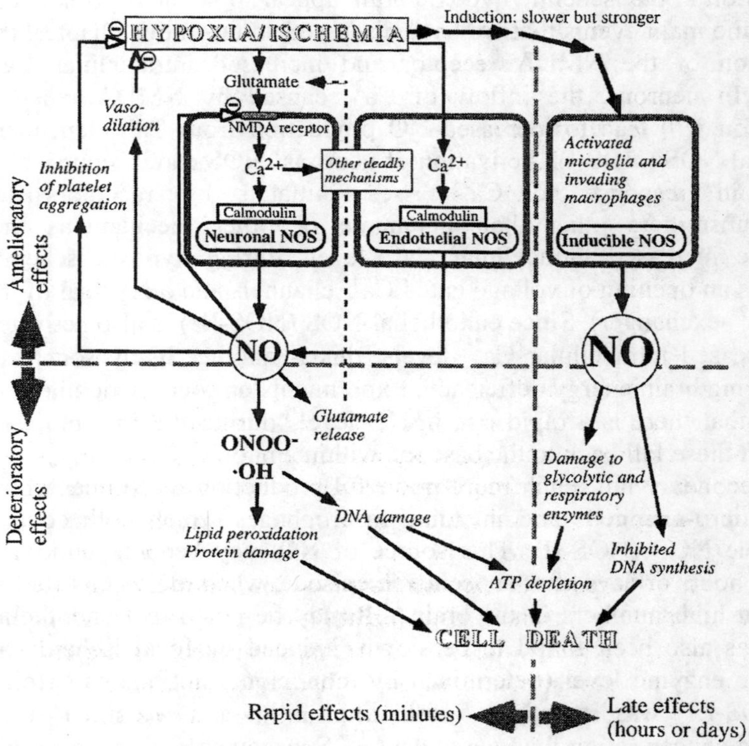

Fig. 4.9 The multitude of detrimental as well as protective effects that NO is likely to have during hypoxia/ischemia. See text for further explanation. Figure inspired by Dawson.[17]

6.1 *Harmful Effects of NO*

NO poses a variety of threats to the neuron. It is a free radical and therefore an unusually reactive biomolecule. The large amounts of NO that can be produced from inducible NOS may represent the greatest danger to the brain since NO may reach concentrations that are high enough to cause direct damage on various cellular components (lower right corner of Fig. 4.9). At such high levels, NO has been found to impair mitochondrial respiration by reacting with key enzymes[17] and it may also impair glycolysis by stimulating an inactivation of glyceraldehyde-3-phosphate dehydrogenase.[137] Moreover, it can inhibit DNA synthesis by reacting with ribonucleotide reductase.[65,41]

From constitutive NOS, only minor amounts of NO can be produced. However, even such small quantities may still be dangerous to the cell (lower left corner of Fig. 4.9) since NO can combine with superoxide (O_2^-) to initiate a production of peroxynitrite ($ONOO^-$) and the hydroxyl radical[4,45] – two highly reactive species that can cause oxidative damage to lipids, proteins and DNA (see also the section on free radicals below). Thus, NO may give rise to molecules that are much more reactive than NO per se. The importance of this mechanism in toxicity mediated by neuronal NOS is indicated by a study showing that the large increase in hydroxyl-like radical formation occurring in response to NMDA infusion in the rat striatum can be completely blocked by NOS inhibitors.[36] The importance of superoxide in the toxicity of NO is further supported by in vitro studies showing that the neurotoxicity of NO donors can be significantly attenuated by the superoxide consuming enzyme superoxide dismutase.[18] Apart from leading directly to cell death, damage to DNA caused by reactive oxidants such as peroxynitrite and the hydroxyl radical (or high levels of NO), leads to the activation of poly(ADP-ribose) synthase – an attempt by the cell to repair DNA. Unfortunately, an over-activated poly(ADP-ribose) synthase system may modify and damage nuclear proteins and is also highly energy consuming, hence, speeding up the ATP depletion during hypoxia/ischemia. In fact, blockers of poly(ADP-ribose) synthase have been found to inhibit NO- and NMDA-mediated neurotoxicity.[138]

Finally, NO can also harm the neuron by stimulating glutamate release[66], which could lead to a vicious circle: glutamate gives more NO which gives more glutamate.

6.2 Beneficial Effects of NO

Since there are studies showing negative effect of NOS inhibitors (e.g. Sancesario et al.[105]) and positive effects of NO-producing substances (e.g. Zhang et al.[136]), NO must also have beneficial effects during cerebral hypoxia/ischemia. The most important of these is probably NO′s ability to cause cerebral vasodilation and, hence, increased perfusion (see the previous chapter). NO has been suggested to be a mediator of the increased blood perfusion seen in the brain during hypoxia[9,50,61], thereby helping to provide more glucose and O_2, as well as remove potentially harmful metabolites such as lactate. In ischemia, a NO-mediated increase in blood flow to the zone surrounding the ischemic territory appears to be a clearly beneficial factor that leads to a reduction of the size of the damaged area.[136]

NO donating drugs and the NO precursor L-arginine inhibit platelet aggregation and adhesion to the arterial wall, while blockers of NOS have the opposite effects (Dawson[17] for review). This suggests that NO stimulates blood flow not only by vasodilation but also by maintaining the fluidity of the blood. However, a study showing that NO donating drugs (SNP and SIN-1) increase brain blood flow and reduce brain damage, indicated that this effect was related to vasodilation, while inhibition of platelet aggregation was of minor or no importance.[136]

Another beneficial effect that NO may have during hypoxia/ischemia is its ability to reduce Ca^{2+} influx through the NMDA receptor[70,81], hence, directly counteracting its glutamate releasing properties.

6.3 NO Conclusions

Although NO appears to produce a complex mixture of deleterious and ameliorating effects during hypoxia/ischemia it appears that they can be separated into early effects, which are generally beneficial, and late effects, which are generally harmful. This is illustrated in Fig 4.9 where harmful effect are denoted with downward directed black arrows, while beneficial effects are shown by upward pointing white arrow. It appears that all of the late (hours or days) effects of the massive production of NO from inducible NOS are deleterious (right part of Fig. 4.9), while at least three of the changes that occur in response to the initial Ca^{2+} mediated NO production favor hypoxic survival. Here, endothelial NOS is probably responsible for most of the beneficial effects, being localized at the best site for increasing brain blood flow without endangering neurons. This view fits well with some experimental data showing that stimulation of NO production at an early stage will be beneficial.[17,136] It should also be emphasized that another factor that complicates the picture is the type of ischemic insult involved: i.e. transient ischemia or permanent ischemia. Dawson[17] suggested that the best strategy to counteract NO related brain damage during ischemia is to stimulate of NO production during the first minutes of ischemia, whereupon a NOS inhibitor could be repeatedly administered at moderate doses over several days. With regard to remedial intervention aimed at blocking harmful NO production during or after hypoxia/ischemia, the isozyme specific NOS inhibitors that our now being developed[27] are likely to prove particularly useful. Indeed, administration of aminoguanidine, a relatively selective inhibitor of inducible NOS, has been found to ameliorate the damage caused by middle cerebral artery occlusion in rats.[46] A 33% reduction of the infarct volume was found even if the drug was administered 24 h post ischemia. The drug tirilazad was suggested to be an inhibitor of neuronal NOS, reducing the effects on brain ischemia in animal models.[23] However, the conclusion from clinical trials with this promising drug, which also acted as a lipid peroxidation inhibitor, is that tirilazad unfortunately worsens the outcome for patients suffering acute ischemic stroke.[127]

However, other more specific inhibitors of neuronal NOS, such as ARL17477 and L-MIN, have been found to reduce anoxic/ischemic brain damage in animal models[87,141], indicating that neuronal NOS is doing more harm than good in this situation. Thus, for the oxygen deprived brain, the only real friend among the NOS isoforms may be the endothelial NOS striving to stimulate cerebral blood supply.

7. FREE RADICAL FORMATION

A free radical is a short lived and chemically unstable molecular species, with at least one of its valence electron orbitals containing a single electron instead of the usual pair needed for stability. Thus, free radicals are often highly reactive due to their inclination to acquire stability by abstracting an electron or a hydrogen atom from molecules they come in contact with. The biologically most important free radicals are the superoxide radical ($*O_2-$), the peroxyhydroxyl radical (HO_2*), and the hydroxyl radical (*OH). In addition, NO is a free radical although it is not as reactive as the molecules mentioned above, and hydrogen peroxide (H_2O_2) is often included among these reactive molecules.

Much of the destruction that occurs following the reestablishment of oxygen supply after ischemia or anoxia has been attributed to the action of free radicals[24,63] and more evidence has since then accumulated in support for this hypothesis.[13,35,107,116] The brain microvasculature appears to be a major target of the free radicals causing dysfunction and disruption of the blood brain barrier.

A major problem is that many of the reactions involving free radicals produce free radicals, so that if unchecked the free radical species will self perpetuate forming a chain reaction of cell injury. The damaging effects are broad and include – denaturation of proteins producing the inactivation of enzymes[24,91], the breakage of DNA strands via base hydroxylation and cross linking, and the initiation of lipid peroxidation, a process probably accelerated by the anoxia induced increase in the free fatty acids levels. Apart from causing a direct destruction of membrane integrity and membrane proteins, lipid peroxidation gives rise to reactive and toxic products such as malondialdehyde. In fact, most research into free radical mechanisms in the CNS has been directed against lipid peroxidation.[108]

Free radicals may also be generated from lipids, for example from arachidonic acid when it is metabolized by cyclooxygenase and lipoxygenase upon reperfusion. Apart from NO and free radicals generated by NO (which we have already discussed), the most important source of free radicals may be the metabolism of the purine metabolites xanthine and hypoxanthine by xanthine oxidase (XO). In anoxia and ischemia, xanthine and hypoxanthine are produced excessively, being break down products of ATP. XO occurs under normal, aerobic, conditions predominantly in a form called xanthine dehydrogenase (XDH) which uses NAD^+ as an electron acceptor, catalyzing the reaction:

$$\text{Xanthine} + H_2O + NAD^+ \rightarrow \text{uric acid} + NADH + H^+$$

During anoxia/ischemia a double jeopardy occurs. Not only does the degradation of ATP lead to an accumulation of xanthine and hypoxanthine, the substrates for superoxide manufacture, but there is also a switch from XDH to XO as the predominant urate producing enzyme. This is thought to be due to the increase in cell Ca^{2+} activating the enzyme calpain which promotes the conversion of XDH to XO. XO uses O_2 as the electron acceptor and produces the superoxide radical ($*O_2$-) by catalyzing the reaction:

$$\text{Xanthine} + H_2O + O_2 \rightarrow \text{uric acid} + 2\ {*O_2}^- + 2H^+$$

Superoxide can subsequently react with hydrogen peroxide in the Haber-Weiss reaction to form the hydroxyl radical (*OH), a reaction catalyzed by Fe^{2+}:

$$*{O_2}^- + H_2O_2 \xrightarrow{Fe^{2+}} OH^- + O_2 + *OH$$

Other possible mechanisms for free radical formation during reperfusion include the dysfunction or over-activity of the respiratory chain resulting in univalent reduction of O_2, neutrophils responding to tissue injury by excessive $*O_2$- formation, or the auto-oxidation of catecholamines.

Since free radicals are produced during normal metabolism, the cell possesses a battery of highly specific defense mechanisms, both enzymatic and non enzymatic, to scavenge free radicals as quickly as they appear. The enzymes involved include two forms of superoxide dismutase (SOD), which transforms $*O_2$- to H_2O_2, as well as catalase and glutathione peroxidase which convert H_2O_2 to O_2 and/or H_2O. In recent experiments utilizing transgenic techniques, it has been shown that an increased level of CuZn-SOD in the brain of transgenic mice protects neurons from ischemic/reperfusion injury, and that a deficiency in CuZn-SOD or mitochondrial Mn-SOD dismutase exacerbates ischemic brain damage.[12] Thus, molecular genetic approaches that modify antioxidant levels in the brain are likely to provide further insight into the roles of oxidants and antioxidant systems in ischemic brain damage.

The antioxidant enzyme systems are more or less closely linked to the non-enzymatic reducing scavenger molecules ascorbate, vitamin E and glutathione. However, during anoxia and ischemia these scavengers may be depleted while free radical production is favored when the supply of oxygen is restored.

Consequently, much experimental effort is going into finding drugs that can function as free radical scavengers, or inhibit the production of free radicals. Some of these approaches have proved to be quite successful in animal models, producing measurable attenuations of reperfusion damage. These include administration of SOD, dimethylthiourea (a *OH scavenger), or allopurinol (a potent inhibitor of XO).[76,82,118]

8. LIPOLYSIS

The tissue levels of free fatty acids rise sharply during anoxia or ischemia.[116] Ca^{2+} appears to play a key role also here by being an activator of lipases, particularly phospholipase A_2 which catalyses the breakdown of phospholipids to free fatty acids. The loss of ATP also plays a role by itself since the build up phospholipids from free fatty acids demands ATP and will therefore stop during energy failure. Thus, the loss of ATP will cause the lipid metabolism to shift towards a net breakdown of phospholipids.

The products of phospholipid hydrolysis during anoxia include free fatty acids (notably arachidonic acid), diacyl glycerol, lysophospholipids and inositol triphosphate (IP_3). Phospholipid hydrolysis can be harmful in a number of ways. First, the breakdown of membrane phospholipids will of course cause a more or less severe disruption of the lipid membrane, thereby possibly opening up the membrane to ion movements.

Secondly, the primary products of phospholipid hydrolysis are likely to threaten neuronal survival. IP_3 is an intracellular messenger that causes further Ca^{2+} release from intracellular stores, thereby adding to the uncontrolled rise in intracellular Ca^{2+}. The free fatty acids and lysophospholipids can act as detergents and ionophores on membranes. Arachidonic acid, as well as some other lipids formed during anoxia/ischemia and reperfusion are potent inhibitors of the Na^+/K^+ pump, and may therefore act to perpetuate the ionic imbalance even if the tissue is reperfused with oxygen.[70]

Thirdly, some of the primary metabolites of phospholipid hydrolysis are further transformed into platelet activating factor (PAF), prostaglandins, leukotrienes and tromboxane A_2, all substances that can cause severe disturbances in key regulatory functions when being produced in excessive amounts.[120] For example, PAF is a powerful proaggregatory and hemostatic agent and it can also initiate inflammatory reactions in blood vessels. Prostaglandin F_{2q} is a very potent inhibitor of the Na^+/K^+ pump.[69]

Finally, if the tissue is reperfused with oxygen, the lipids that have accumulated during anoxia/ischemia may be excessively oxidized, leading to the formation of highly reactive and cytotoxic aldehydes and free radicals (see above).

9. CELL SWELLING

Another important process that is likely to cause neuronal death is cell swelling.[73] Normally, the interstitial space in brain is about 20 % of the brain volume. Cell swelling starts during phase 1, and there is a drastic and sudden fall in the brain interstitial space down to 6% of the brain volume during anoxic depolarization.[37] Cell swelling during anoxia or ischemia appears to be caused by an obligatory inflow of water accompanying the influx of ions, as illustrated by the fact that glutamate induced cell swelling in cultured neurons can be prevented if Na^+ or Cl^- are removed from the culture medium.[104] For astrocytes, which make up about 25% of the cellular volume of the brain, an active accumulation of lactate at acidic pH is thought to be an important mechanism of postischemic swelling.[77]

One of the reasons why a partial reduction in cerebral blood flow can cause more damage than complete cerebral ischemia, is that a complete halt in blood flow will limit cell swelling because no additional water is transported to the brain to continue the swelling process. However, experiments with cultured neurons have showed that cell swelling *per se* does not have to be incompatible with neuronal survival. Thus, Choi[14] showed that mammalian neocortical neurons could tolerate 5 min periods of cell swelling caused by glutamate exposure as long as Ca^{2+} was eliminated from the culture medium, while if Ca^{2+} was added, the neurons eventually died. The major problem with cell swelling in the vertebrate brain is probably related to the fact that the brain is confined in a hard bone envelope. Thus, cell swelling will result in an increased cerebral pressure, and when this pressure exceeds the blood pressure, perfusion to the brain will be stopped even if the original cause of the ischemia or anoxia is eliminated.

10. LACTATE AND ACIDOSIS

The final step of anaerobic glycolysis – the lactate dehydrogenase reaction – is chiefly responsible for the regeneration of the NAD consumed in one of the central steps of glycolysis (the glyceraldehyde-

3-phosphate reaction). Thus, if glycolysis would end at pyruvate during anaerobiosis, NAD would soon be depleted and glycolysis would stop. Unfortunately, the result is an accumulation of lactate as a metabolic end product, and intracellular acidosis. Since changes in intracellular pH have widespread metabolic effects[10] excessive acidosis is thought to be a major mechanism causing brain damage.[19,92] In sustained ischemia the harmful effects that have been linked to severe neuronal acidosis include edema, inhibition of mitochondrial respiration, inhibition of lactate oxidation, the inhibition of H^+ extrusion and production of free radicals.[118] It should be noted however that cellular acidification is correlated with, but not actually caused by, lactate build-up, and indeed the relationships between changes in intracellular pH and lactate can be quite different in different physiological states.[8,92] For example, during anaerobiosis, ATP hydrolysis is a major source of H^+ production, and at physiological pHs the hydrolysis of creatine phosphate causes H^+ to be taken up on a mol per mole basis, the reaction acting as a pH buffer.[43] In astrocytes there is evidence of a facilitated cotransport of lactate with H^+ or exchange for a hydroxyl ion with the consequent accumulation of extracellular lactate at acid pH.[77] For example at a media pH of 6.0 with 20mmol lactate the intracellular concentration of lactate in a cell culture was about 60mmol and this osmotic gradient is thought to be a significant factor in inducing cell swelling.[77]

It is well established that hyperglycemia exacerbates ischemic damage, which has been attributed to the enhanced production of lactic acid in the anoxic brain resulting from the increased substrate available for anaerobic glycolysis.[94,99] However, in some circumstances hyperglycemia may actually be beneficial, particularly in anoxia where lactate can be flushed away by the continuing blood flow, as against stagnant ischemia where lactate must accumulate. In the former case hyperglycemia may actually facilitate ATP synthesis without causing harmful acidosis. It has been shown that perfusing a high concentration of glucose (20 mM) via a microdialysis probe to the ischemic rat caudate nuclei greatly reduced the release of excitotoxic glutamate compared to glucose free perfusates.[124] As far as the incubated brain slice is concerned hypoglycemia worsens and hyperglycemia assists recovery of neuronal function from hypoxia.[111] Indeed, there is evidence that acidosis is not a factor in the selective neuronal necrosis that occurs after transient ischemia and that moderate acidosis may actually play a protective role during ischemia [128]. In fact, Schurr et al.[103] reported that in the hypoxic rat hippocampal slice 10mM lactate increased the recovery rate of synaptic function and 20 mM lactate with pH values as low as 5.5 had no detrimental effects. For a short period when oxygen is restored after anoxia/ischemia, and while ATP is still depleted, lactate may even be preferred over glucose as a fuel, since glucose requires ATP to enter into the glycolytic pathway while lactate is converted to pyruvate at no ATP cost.[111] Indeed, rat hippocampal slices can maintain neuronal function with lactate as the sole fuel.[111] Thus, it is possible that hyperglycemic damage may be due to mechanisms other than lactate acidosis.

11. MITOCHONDRIAL DAMAGE

As mentioned, neurons often start or continue to die long after an anoxic/ischemic insult is over. This progressive late brain damage may be related to changes in postischemic metabolism. Interestingly, the metabolic rate of the brain can remain reduced for a day or more after an ischemic insult. Pulsinelli et al. [97] found that rats subjected to 30 min of forebrain ischemia displayed depressed rates of glucose utilization in striatum, neocortex and the hippocampus for 48 h post ischemia. They also noted that in the hippocampus, glucose utilization equaled the control value after 1 h post ischemia, whereupon it fell below control between 24 and 48 h. It is likely that such alterations in metabolic rate are related to delayed damage of the mitochondria – the major ATP producers in the cell.

One factor that may be especially important for delayed mitochondrial dysfunction is damage to the mitochondrial DNA. Abe et al.[1] have summarized evidence suggesting that that expression of mitochondrial genes is more sensitive to an ischemic insult than that of nuclear genes, at least in the CA1 region of the hippocampus. For example, three of the 13 subunits of cytochrome c oxidase are encoded by mitochondrial DNA and the activity of this enzyme shows a striking progressive fall in the activity in the CA1 region, from 1 h to 7 days, a fall that starts before neuronal death is apparent.

However, mitochondrial dysfunction may not be a general phenomenon in the post ischemic brain. Thus, Katsura et al.[57] have presented data indicating that the metabolic capacity of the cortex in the post-ischemic rat brain is intact. They found that although the basal level of oxygen consumption was depressed by about 50% for at least 6 h after the ischemic insult, they could provoke the post-ischemic cortex (through bicuculline induced seizures) to upregulate its oxygen consumption to a rate that was equal to that in an identically provoked control brain.

12. APOPTOSIS

It is now generally accepted that programmed cell death, or apoptosis, can be triggered by ischemia. A number of studies have identified characteristic molecular indicators associated with programmed cell death in the ischemic brain. These include the translocation of cytochrome c from mitochondria to cytosol and the activation of caspases.[12,78] In brain, caspase-3 appears to be of particular importance. Present evidence suggests that apoptosis is an especially important mediator of delayed cell death within the ischemic penumbra, where the activated caspases break down the cell by cleaving cytoskeletal proteins, DNA repair enzymes and other regulatory and structural proteins, and cell repair proteins.

Experimental inhibition of caspases in mammalian models of ischemia has been found to reduce the damage, and the therapeutic window for this intervention appears to be longer than for blocking glutamate receptors. Indeed, combining both treatments (caspase inhibition and glutamate receptor blockade) seems to expand the treatment window synergistically.[100,109]

Pathways of programmed cell death in the focal and global ischemia models are examined in more detail in the following chapter

13. CONCLUSIONS

As we have seen anoxia rapidly produces an almost chaotic disruption in a plethora of neuronal processes. Although some may turn out to be of minor importance as far as causing actual cell damage, many of them are linked. But the rapidity and complexity of the degenerative events that occur in the anoxic brain renders it extremely difficult to sort out cause and effect sequences or to identify the most vulnerable processes (the Achilles heel) that initiates the final collapse. Consequently, protective or curative measures against ischemic or anoxic damage, that attack only secondary or tertiary detrimental mechanisms, are likely to be no more than partially successful.

Bearing in mind the numerous obstacles for research to enhance anoxic brain survival in mammals, a fresh perspective can be obtained from those exotic vertebrates that have the unique ability to survive prolonged anoxia. In contrast to modern medical science, evolution solved the problems of anoxic brain survival millions of years ago.

14. REFERENCES

1. Abe K, Aoki M, Kawagoe J, Yoshida T, Hattori A, Kogure K, Itoyama Y. Ischemic delayed neuronal death, a mitochondrial hypothesis. Stroke 1995; 26:1478-1489.
2. Andine P, Orwar O, Jacobson I, Sandberg M, Hagberg H. Changes in extracellular amino acids and spontaneous neuronal activity during ischemia and extended reflow in the CA1 of the rat hippocampus. J Neurochem 1991; 57:222-229.
3. Beal MF, Hyman BT, Koroshetz W. Do defects in mitochondrial energy metabolism underlie the pathology of neurodegenerative diseases? Trends Neurosci 1993; 16:125-131.
4. Beckman JS, Crow JP. Pathological implications of nitric oxide, superoxide and peroxynitrite formation. Biochem Soc Trans 1993; 21:330-334.
5. Bertorello AM, Hopfield JF, Aperia A, Greengard P. Inhibition by dopamine of (Na^+-K^+) ATPase activity through D_1 and D_2 dopamine receptor synergisms. Nature 1990; 347:386-388.
6. Berndt C, Henke W, GrossJ. Hypoxia induces different responses of striatal high and low affinity dopamine uptake sites. Mol and Chem Neuropath 1993; 18:179-187.
7. Broderick PA, Gibson GE. Dopamine and serotonin in rat striatum during in vivo hypoxic-hypoxia. Met Brain Dis 1989; 4:143-153.
8. Brooks KJ, Porteous R, Bachelard HS. Effects of hypoglycemia and hypoxia on the intracellular pH of cerebral tissue as measured by ^{31}P nuclear magnetic resonance. J Neurochem 1989; 52:604-610.
9. Burnstock G. Hypoxia, endothelium, and purines. Drug Development Res 1993; 28:301-305
10. Carafoli E: Intracellular calcium homeostasis. Annu Rev Biochem 1987; 56:395-433.
11. Cazevieille C, Muller A, Meynier F, Bonne C. Superoxide and nitric oxide cooperation in hypoxia/reoxygenation-induced neuron injury. Free Radical Biol Med 1993; 14:389-395.
12. Chalmers-Redman RM, Fraser AD, Ju WY, Wadia J, Tatton NA, Tatton WG. Mechanisms of nerve cell death: apoptosis or necrosis after cerebral ischaemia. Int Rev Neurobiol. 1997; 40:1-25.
13. Chan PH. Role of oxidants in ischemic brain damage. Stroke 1996; 27:1124-9.
14. Choi D. Ionic dependence of glutamate neurotoxicity in cortical cell culture. J Neurosci 1987; 7:380-390.
15. Choi DW: Calcium: still center stage in hypoxic-ischemic neuronal death. Trends Neurosci 1995; 18:58-60.
16. Czéh G, Aitken PG, Somjen GG. Whole-cell membrane current and membrane resistance during hypoxic spreading depression. Neuro Report 1992; 3:197-200.
17. Dawson DA: Nitric oxide and focal cerebral ischemia: multiplicity of actions and diverse outcome. Cerebrovasc Brain Metab Rev 1994; 6:299-324.
18. Dawson VL, Dawson TM, Bartley DA, Uhl GR, Snyder SH. Mechanisms of nitric oxide-mediated neurotoxicity in primary brain cultures. J Neurosci 1993; 13:2651-2661.
19. DeCourten-Meyers G, Myers RE, Schoolfield L. Hyperglycemia enlarges infarct size in cerebrovascular occlusion in cats. Stroke 1988; 19:623-630.
20. Ekholm A, Asplund B, Siesjö BK. Perturbation of cellular energy state in complete ischemia: relationship to dissipative ion fluxes. Exp Brain Res 1992; 90:47-53.
21. Ekholm A, Katsura K, Siesjö BK. Coupling of energy failure and dissipative K^+ flux during ischemia: role of preischemic plasma glucose concentration. J Cereb Blood Flow Metab

1993; 13:193-200.
22. Erecinska M, Silver IA. ATP and brain function. J Cereb Blood Flow Metab 1989; 9:2-19.
23. Fernandez MP, Belmonte MJ, Meizoso MJ, Garcia-Novio M. Desmethyl tirilazad reduces brain nitric oxide synthase activity and cyclic guanosine monophosphate during cerebral global ischemia in rats. Res. Commun. Mol. Pathol. Pharmacol. 1997; 95:33-42.
24. Flamm ES, Demopoulos HB, Seligman ML, Poser RG, Ransohoff J. Free radicals in cerebral ischemia. Stroke 1978; 9:445-447.
25. Folbergrova J, Kiyota Y, Pahlmark K, Memezawa H, Smith ML, Siesjo BK. Does ischemia with reperfusion lead to oxidative damage to proteins in the brain. J Cereb Blood Flow Metab 1993; 13:145-152.
26. Fujisawa H, Dawson D, Browne SE, MacKay KB, Bullock R, McCulloch J. Pharmacological modification of glutamate neurotoxicity in vivo. Brain Res 1993; 629:73-78.
27. Fu*kuto JM, Chaudhuri G. Inhibition of constitutive and inducible nitric oxide synthase: potential selective inhibition. Ann Rev Pharmacol Toxicol 1995; 35:165-194.
28 Globus MY, Busto R, Dietrich WD, Martinez E, Valdes I, Ginsberg MD. Effect of ischemia on the in vivo release of striatal dopamine, glutamate and □-aminobutyric acid studied by intracerebral microdialysis. J Neurochem 1988; 51:1455-1464.
29. Globus MY, Busto R, Dietrich WD, Martinez E, Valdes I, Ginsberg MD. Direct evidence for acute and massive norepinephrine release in the hippocampus during transient ischemia. J Cereb Blood Flow Metab 1989; 9:892-896.
30. Globus MY-T, Ginsberg MD, Busto R. Excitotoxic index-a biochemical marker of selective vulnerability. Neurosci Letters 1991; 127:39-42.
31. Globus MY, Busto R, Martinez E, Valdes I, Dietrich WD, Ginsberg MD. Comparative effects of transient global ischemia on extracellular levels of glutamate, glycine, and □-aminobutyric acid in vulnerable and nonvulnerable brain regions in the rat. J Neurochem 1991; 57:470-478.
32. Globus MY-T, Prado R, Sanchez-Ramos J, Zhao W, Deitrich WD, Busto R, Ginsberg MD. A dual role for nitric oxide in NMDA-mediated toxicity in vivo. J Cereb Blood Flow Metab 1995; 15:904-913.
33. Grigg JJ, Anderson EG. Glucose and sulfonyureas modify different phases of the membrane potential changes during hypoxia in rat hippocampal slices. Brain Res 1989; 409:302-310.
34. Hagberg H, Lehmann A, Sandberg M, Nystrom B, Jacobson I, Hamberger A. Ischemia-induced shift of inhibitory and excitatory amino acids from intra- to extracellular compartments. J Cereb Blood Flow Metab 1985; 5:413-419.
35. Halliwell B, Gutteridge J. Free Radicals in Biology and Medicine. 2nd ed. Oxford: Clarendon Press,1989.
36. Hammer B, Parker WD, Bennett JP. NMDA receptors increase OH radicals in vivo by using nitric oxide synthase and protein kinase C NeuroReport 1993; 5:72-74.
37. Hansen AJ. The extracellular potassium concentration in brain cortex following ischemia in hypo- and hyperglycemic rats. Acta Physiol Scand 1978; 102:324-329.
38. Hansen AJ. Effect of anoxia on ion distribution in the brain. Physiol Rev 1985; 65:101-148.
39. Hansen AJ, Olsen CE. Brain extracellular space during spreading depression and ischemia. Acta Physiol Scand 1980; 108:355-365.
40. Hashimoto N, Matsumoto T, Mabe H, Hashitani T, Nishino H. Dopamine has inhibitory and accelerating effects on ischemia-induced neuronal cell damage in the rat striatum. Brain Res Bull 1994; 33:281-288.
41. Henry Y, Lepoivre M, Drapier JC, Ducrocq C, Boucher JL, Guissani A. EPR characterization of molecular targets for NO in mammalian cells and organelles. FASEB J 1993; 7:1124-1134.
42. Hillered L, Chan PH. Role of archidonic acid and other fatty acids in mitochondrial dysfunction in brain ischemia. J Neurosci Res 1988; 20:451-456.
43. Hochachka PW, Mommsen TP. Protons and anaerobiosis. Science 1983; 219:1391-1397.
44. Hochachka PW, Somero GN. Biochemical Adaptation. Princeton: Princeton University

Press,1984.
45. Hogg N, Darley-Usmar VM, Wilson MT, Moncada S. Production of hydroxyl radicals from the simultaneous generation of superoxide and nitric oxide. Biochem J 1992; 281:419-424.
46. Huang C-C, Lajevardi NS, Tammela O, Pastuszko A, Delivoria-Papadopoulos M, Wilson DF. Relationship of extracellular dopamine in striatum of newborn piglets to cortical oxygen pressure. Neurochem Res 1994; 19: 649-655.
47. Hylland P, Nilsson GE, Johansson D. Anoxic brain failure in an ectothermic vertebrate: release of amino acids and K^+ in rainbow trout thalamus. Am J Physiol 1995; 269: R1077-R1084.
47. Iadecola C, Zhang F, Xu X. Inhibition of inducible nitric oxide synthase ameliorates cerebral ischemic damage. Am J Physiol 1995; 268:R286-R292.
49. Ikeda M, Nakazawa T, Abe K, Kaneko T, Yamatsu K. Extracellular accumulation of glutamate in the hippocampus induced by ischemia is not calcium dependant- in vitro and in vivo evidence. Neurosci Lett 1989; 96:202-206.
50. Isozumi K, Fukuuchi Y, Takeda H, Itoh Y. Mechanisms of CBF augmentation during hypoxia in cats: probable participation of prostacyclin, nitric oxide and adenosine. Keio J Med 1994; 43:31-36.
51. Jiang C, Xia Y, Haddad GG. Role of ATP-sensitive K^+ channels during anoxia: major differences between rat (newborn and adult) and turtle neurones. J Physiol Lond 1992; 448:599-612.
52. Jing J, Aitken PG, Somjen GG. Role of calcium channels in spreading depression in rat hippocampal slices. Brain Res 1993; 604:251-259.
53. Jonas P, Koh DS, Kampe K, Hermsteiner M, Vogel W. ATP-sensitive and Ca-activated K channels in vertebrate axons: novel links between metabolism and excitability. Pflugers Arch 1991; 418:68-73.
54. Kader A, Frazzini VI, Solomon RA, Trifiletti RR. Nitric oxide production during focal cerebral ischemia in rats. Stroke 1993; 24:1709-1716.
55. Katayama Y, Kawamata T, Tamura T, Hovda DA, Becker DP, Tsubokawa T. Calcium-dependent glutamate release concommitant with massive potassium flux during cerebral ischemia in vivo. Brain Res 1991; 558:136-140.
56. Katchman AN, Hershkowitz N. Early anoxia-induced vesicular glutamate release results from mobolization of calcium from intracellular stores. J Neurophysiol 1993; 70:1-7.
57. Katsura K, Kristián T, Siesjö BK. Energy metabolism, ion homeostasis, and cell damage in the brain. Biochem Soc Trans 1994; 22:991-996.
58. Kim D, Sladek CD, Aguado-Velasco C, Mathiasen JR. Arachidonic acid activation of a new family of K^+ channels in cultured rat neuronal cells. J Physiol 1995; 484:643-660.
59. Koike T, Martin DP, Johnson EM Jr. Role of Ca^{2+} channels in the ability of membrane depolarization to prevent neuronal death induced by trophic-factor deprivation: evidence that levels of internal Ca^{2+} determine nerve growth factor dependence of sympathetic ganglion cells. Proc Natl Acad Sci USA 1989; 86: 6421-6425.
60. Korf J, Klein HC, Venema K, Postema F. Increases in striatal and hippocampal impedance and extracellular levels of amino acids by cardiac arrest in freely moving rats. J Neurochem 1988; 50:1087-1096.
61. Kozniewska E, Oseka M, Stys T. Effects of endothelium-derived nitric oxide on cerebral circulation during normoxia and hypoxia in the rat. J Cereb Blood Flow Metab 1992; 12:311-317.
62. Krnjevic MB. Adenosine triphoshpate-sensitive potassium channels in anoxia. Stroke 1990; Supplement III. 21:190-193.
63. Krnjevic K. Membrane current activation during hypoxia in hippocampal neurones. In: Surviving Hypoxia: Mechanisms of Control and Adaptation. P. W. Hochachka, P. L. Lutz, T. Sick, M. Rosenthal and G.van den Thilart (Eds). Boca Raton, FL: CRC Press, 1993: 365-388.
64. Kumagae Y, Matsui Y. Output, tissue levels, and synthesis of acetylcholine during and after transient forebrain ischemia in the rat. J Neurochem 1991; 56:1169-1173.

65. Kwon NS, Stuehr DJ, Nathan CF. Inhibition of tumor cell ribonucleotide reductase by macrophage-derived nitric oxide. J Exp Med 1991; 174:761-767.
66. Lawrence AJ, Jarrott, B. Nitric oxide increases interstitial excitatory amino acid release in the rat dorsomedial medulla oblongata. Neurosci Lett 1993; 151:126-129.
67. Leblond J, Krnjevic K. Hypoxic changes in hippocampal neurones. J Neurophysiol 1989; 62:1-14.
68. Lee JM, Grabb MC, Zipfel GJ, Choi DW. Brain tissue responses to ischemia. J Clin Invest. 2000; 106:723-31.
69. Lees GJ. Inhibition of sodium-potassium-ATPase: a potentially ubiquitous mechanism contributing to central nervous system neuropathology. Brain Res Rev 1991; 16:283-300.
70. Lei SZ, Pan Z-H, Aggarwal SK, Chen H-SV, Hartman J, Sucher NJ, Lipton SA. Effect of nitric oxide production on the redox modulatory site of the NMDA receptor-channel complex. Neuron 1992; 8:1087-1099.
71. Linnik MD, Zobrist RH, Hatfield MD. Evidence supporting a role for programmed cell death in focal cerebral ischemia in rats. Stroke. 1993; 24:2002-2008.
72. Lipton SA, Kater SB. Neurotransmitter regulation of neuronal outgrowth, plasticity and survival. Trends Neurosci 1989; 12:265-270.
73. Lipton, P. Ischemic Cell Death in Brain Neurons. Physiol. Rev. 1999. 79: 1431–1568, .
74. Lipton P, Whittingham TS. Reduced ATP concentrations as a basis for synaptic transmission failure during hypoxia in the in vitro guinea-pig hippocampus. J Physiol 1982; 325:51-56.
75. Lipton P, Whittingham TS. Energy metabolism and brain slice function. In: Dingledine D ed(s). Brain Slices. New York: Plenum,1984: 113-153.
76. Liu TH, Beckman JS, Freeman BA, Hogan EL, Hsu CY. Polyethylene glycol-conjugated superoxide dismutase and catalase reduce ischemic brain injury. Am J Physiol 1989; 256:H589-H593.
77. Lomneth RS, Medrano EI, Gruenstein. The role of transmembrane pH gradients in the lactic acid induced swelling of astroctytes. Brain Res 1992; 523: 69-77.
78. MacManus JP, Linnik MD.Gene expression induced by cerebral ischemia: an apoptotic perspective. J Cereb Blood Flow Metab. 1997; 17:815-32
79. Madl JE, Burgesser K. Adenosine triphosphate depletion reverses sodium-dependent, neuronal uptake of glutamate in rat hippocampal slices. J Neurosci 1993; 13:4429-2222.
80. Malinski T, Bailey F, Zhang ZG, Chopp M. Nitric oxide measured by a porphyrinic microsensor in rat brain after transient middle cerebral artery occlusion. J Cereb Blood Flow Metab 1993; 13:355-358.
81. Manzoni O, Prezeau I, Martin P, Deshager S, Bockaert J, Fagni L. Nitric-oxide induced blockade of NMDA receptors. Neuron 1992; 8:653-662.
82. Martz D, Rayos G, Schielke GP, Betz AL. Allopurinol and dimethylthiourea reduce brain infarction following middle cerebral artery occlusion in the rat. Stroke 1989; 20:488-494.
83. Matsumoto K, Ueda S, Hashimoto T, Kuriyama K. Ischemic neuronal injury in the rat hippocampus following transient forebrain ischemia: evaluation using in vivo microdialysis. Brain Res 1991; 543:236-242.
84. Moghaddam B, Schenk JO, Stewart WB, Hansen AJ. Temporal relationship between neurotransmitter release and ion flux during spreading depression and anoxia. Can J Physiol Pharmacol 1987; 65:1105-1110.
85. Mourre C, Ben Ari Y, Bernardi H, Fosset M, Lazdunski M. Antidiabetic sulfonylurea: location of binding sites in the brain and effects on the hyperpolarization induced by anoxia in hippocampal slices. Brain Res 1989; 486:159-164.
86. Nellgard B, Wieloch T. Postischemic blockade of AMPA but not NMDA receptors mitigates neuronal damage in the rat brain following transient severe cerebral ischemia. J Cereb Blood Flow Metab 1992; 12:2-11.
87. Nagafuji T, Sugiyama M, Muto A, Makino T, Miyauchi T, Nabata H. The neuroprotective effect of a potent and selective inhibitor of type I NOS (L-MIN) in a rat model of focal cerebral ischaemia. Neuroreport. 1995; 6:1541-1545.

88. Nilsson GE, Perez-Pinzon M, Dimberg K, Winberg S. Brain sensitivity to anoxia in fish as reflected by changes in extracellular K^+ activity. Am J Physiol 1993; 264:R250-R253.
89. Nowicki JP, Duval D, Poignet H, Scatton B. Nitric oxide mediates neuronal death after focal cerebral ischemia in the mouse. Eur J Pharmacol 1991; 204:339-340.
90. Obrenovitch TP, Richards DA. Extracellular neurotransmitter changes in cerebral ischaemia. Cerebrovasc Brain Metab Rev 1995; 7:1-54.
91. Oliver CN, Starke-Reed PE, Stadtman ER, Liu GJ, Carney JM, Floyd RA. Oxidative damage to brain proteins, loss of glutamine synthetase activity, and production of free radicals during ischemia/reperfusion-induced injury to gerbil brain. Proc Natl Acad Sci USA 1990; 87:5144-5147.
92. Paschen W, Djuricic B, Mies G, Schmidt-Kastner R, Linn F. Lactate and pH in the brain: association and dissociation in different pathophysiological states. J Neurochem 1987; 48:154-159.
93. Pérez-Pinzón MA, Nilsson G, Lutz PL. Relationship between ion gradients and neurotransmitter release in the newborn rat striatum during anoxia. Brain Res 1993; 602:228-233.
94. Phillis JW, Simpson RE, Walter GA. The effect of hyperglycemia on extracellular levels of adenosine in the hypoxic rat cerebral cortex. Brain Res 1990; 524:336-338.
95. Phillis JW, Walter GA, Simpson RE. Brain adenosine and transmitter amino acid release from the ischemic rat cerebral cortex: effects of the adenosine deaminase inhibitor deoxycoformycin. J Neurochem 1991; 56:644-650.
96. Pulsinelli WA. Selective neuronal vulnerability: morphological and molecular characteristics. In: Kogure K, Hossman KA, Siesjö BK, (eds). Molecular Mechanisms of Ischemic Brain Damage: Progress in Brain Research. Amsterdam: Elsevier Science,1985: 29-37.
97. Pulsinelli WA, Levy DE, Duffy TE. Regional cerebral blood flow and glucose metabolism following transient forebrain ischemia. Ann Neurol 1982; 11:499-502.
98. Puro DG. Stretch-activated channels in human retinal Muller cells. Glia 1991; 4:456-460.
99. Rehncrona S, Rosen I, Siesjö BK. Excessive cellular acidosis: an important mechanism of neuronal damage in brain? Acta Physiol Scand 1980; 110:435-437.
100. Robertson, GS, Crocker, SJ, Nicholson, DW and Schulz, JB. Neuroprotection by the inhibition of apoptosis. Brain Pathol. 2000; 10: 283-292.
101. Rossen R, Kabat H, Anderson JP. Acute arrest of cerebral circulation in man. Arch Neurol Psychiatry 1943; 50:510-528.
102. Rothman SM, Olney JW. Excitotoxicity and the NMDA receptor – still lethal after eight years. Trends Neurosci 1995; 18:57-58.
103. Rothman SM, Thurston JG, Hauhart RE. Delayed neurotoxicity of excitatory amino acids in vitro. Neurosci 1987; 22:471-480.
104. Rothman SM: Excitotoxins: possible mechanisms of action. Ann. New York Acad Sci 1992; 648:132-139.
105. Sancesario G, Iannone M, Morello M, Nisticò G, Bernardi G. Nitric oxide inhibition aggavates ischemic damage of hippocampal but not of NADPH neurons in gerbils. Stroke 1994; 25:436-444.
106. Sánchez-Prieto J, González P. Occurence of a large Ca^{2+}-independent release of glutamate during anoxia in isolated nerve terminals. J Neurochem 1988; 50:1322-1324.
107. Sarna GS, Obrenovitch TP, Matsumoto T, Symon L, Curzon G. Effect of transient cerebral ischaemia and cardiac arrest on brain extracellular dopamine and serotonin as determined by in vivo dialysis in the rat. J Neurochem 1990; 55:937-940.
108. Schmidley JW. Free radicals in central nervous system ischemia. Stroke 1990; 21:1086-1090.
109. Schulz JB, Weller M, Moskowitz MA. Caspases as treatment targets in stroke and neurodegenerative diseases. Ann Neurol 1999; 45: 421-429..
110. Schurr A, Dong WQ, Reid KH, West CA, Rigor BM. Lactic acidosis and recovery of neuronal function following cerebral hypoxia in vitro. Brain Res 1988; 438:311-314.

111. Schurr A, Rigor BM: Cerebral ischemia revisited: new insights as revealed using in vitro brain slice preparation. Experientia 1989; 45:684-695.
112. Shigeno T, Yamasaki Y, Kato G, Kusaka K, Mima T, Takakura K, Graham DI, Furukawa S. Reduction of delayed neuronal death by inhibition of protein synthesis. Neurosci Lett 1990; 120: 117-119.
113. Sick TJ, Rosenthal M, LaManna JC, Lutz PL. Brain potassium ion homeostasis, anoxia, and metabolic inhibition in turtles and rats. Am J Physiol 1982; 243:R281-R288.
114. Siesjö BK. Brain Energy Metabolism. New York: John Wiley,1978.
115. Siesjö BK. Calcium, excitotoxins, and brain damage. News Physiol Sci 1990; 5:120-125.
116. Siesjö BK, Katsura K. Ischemic brain damage: focus on lipids and lipid mediators. Adv Exp Med Biol 1992; 318:41-56.
117. Siesjo BK, Zhao Q, Pahlmark K, Siesjo P, Katsura K, Folbergrova J. Glutamate, calcium, and free radicals as mediators of ischemic brain damage. Ann Thorac Surg 1995; 59:1316-20.
118. Siesjö BK. Pathophysiology and treatment of focal cerebral ischemia. Part I: Pathophysiology. J Neurosurg 1992; 77:169-184.
119. Siesjö BK. Calcium-mediated processes in neuronal degeneration. Ann NY Acad Sci 1994; 747:140-161.
120. Siesjö BK, Katsura K, Kristián T. The biochemical basis of cerebral ischemic damage. J Neurosurgical Anesthesiology 1995; 7:47-52.
121. Silver IA, Erecinska M. Intracellular and extracellular changes of [Ca^{2+}] in hypoxia and ischemia in rat brain in vivo. J Gen Physiol 1990; 95:837-865.
122. Somjen GG, Aitken PG, Czeh GL, Herreras O, Jing J, Young JN. Mechanisms of spreading depression: a review of recent findings and a hypothesis. Can J Physiol Pharmacol 1992; 70:S248-S254.
123. Somjen GG, Aitken PG, Czeh G, Jing J, Young JN. Cellular physiology of hypoxia of the mammalian central nervuos system. In: Waxman SG ed(s). Molecular and cellular approaches in treatment of neurological diseases. New York: Raven Press,1993: 51-65.
124. Swanson RA. Glucose can fuel glutamate uptake in ischemic brain. J Cerebral Blood Flow Metab 1994; 14:1-6.
125. Szatkowski M, Attwell D. Triggering and execution of neuronal death in brain ischaemia: two phases of glutamate release by different mechanisms. Trends Neurosci 1994; 17:359-65.
126. Taylor CP, Geer JJ, Burke SP. Endogenous extracellular glutamate accumulation in rat neocortical cultures by reversal of the transmembrane sodium gradient. Neurosci Lett 1992; 145:197-200.
127. Tirilazad International Steering Committee. Tirilazad mesylate in acute ischemic stroke, a systematic review. Stroke 2000; 31:2257-2265
128. Tombaugh GC, Sapolsky RM. Evolving concepts about the role of acidosis in ischemia neuropathology. J Neurochem 1993; 61:793-803.
129. Ungerstedt U. Measurement of neurotransmitter release by intracranial dialysis. In: Marsden CA ed(s). Measurement of Neurotransmitter Release In Vivo. New York: Wiley,1984: 81-105.
130. Valentino K, Newcomb R, Gadbois T. A selective N-type calciumchannel antagonist protects against neuronal loss after global ischemia. Proc Natl Acad Sci USA 1993; 90:7894-7897.
131. Xia Y, Jaing C, Haddad GG. Oxidative and glycolytic pathways in rat (newborn and adult) and turtle brain: role during anoxia. Am J Physiol 1992; 262:R595-R603.
132. Xie Y, Dengler K, Zacharias E, Wilffert B, Tegtmeier F. Effects of the sodium channel blocker tetrodotoxin (TTX) on cellular ion homeostasis in rat brain subjected to complete ischemia. Brain Research 1994; 652, 216-224.
133. Yamamoto S, Golanov EV, Berger SB, Reis DJ. Inhibition of nitric oxide synthesis increases focal ischemic infarction in rat. J Cereb Blood Flow Metab 1992; 12:717-726.
134. Zeevalk GD, Nicklas WJ. Evidence that the loss of the voltage-dependent Mg^{2+} block at the N-methyl-D-aspartate receptor underlies receptor activation during inhibition of neuronal

metabolism. J Neurochem 1992; 59:1211-1220.
135. Zhang F, Iadecola C. Nitroprusside improves blood flow and reduces brain damage after focal ischemia. Neuro Report 1993; 4:559-562.
136. Zhang F, White JG, Iadecola C. Nitric oxide donors increase blood flow and reduc brain damage in focal ischemia: evidence that nitric oxide is beneficial in early stages of cerebral ischemia. J Cereb Blood Flow Metab 1994a; 14:217-226.
137. Zhang J, Snyder SH. Nitric oxide stimulates auto-ADP-ribosylation of glyceraldehyde-3-phosphate dehydrogenase. Proc Natl Acad Sci USA 1992; 89:9382-9385.
138. Zhang J, Dawson VL, Dawson TM, Snyder SH. Nitric oxide activation of poly poly(ADP-ribose) synthetase in neurotoxicity. Science 1994b; 263:687-689.
139. Zhang ZG, Chopp M, Zaloga C, Pollock JS, Förstermann U. Cerebral endothelial nitric oxide synthase expression after focal cerebral ischemia in rats. Stroke 1993; 24:2016-2022.
140. Zhang ZG, Chopp M, Gautam S, Zaloga C, Zhang RL, Schmidt HH, Pollock JS, Forstermann U. Upregulation of neuronal nitric oxide synthase and mRNA, and selective sparing of nitric oxide synthase-containing neurons after focal cerebral ischemia in rat. Brain. Res. 1994c; 654:85-95.
141. Zhang ZG, Reif D, Macdonald J, Tang WX, Kamp DK, Gentile RJ, Shakespeare WC, Murray RJ, Chopp M. ARL 17477, a potent and selective neuronal NOS inhibitor decreases infarct volume after transient middle cerebral artery occlusion in rats. J Cereb Blood Flow Metab. 1996; 16:599-604.

CHAPTER 5

MOLECULAR ASPECTS OF BRAIN ISCHEMIA IN MAMMALS

1. INTRODUCTION

Neural molecular adaptations to hypoxia/ischemia include a variety of changes in expression of regulatory proteins implicated in promoting or antagonizing processes of delayed neuronal death. In particular the early transcriptional induction of key transcription factor activities may be accompanied by the triggering of pathways for survival including elevations in stress and heat shock proteins. As we have seen in chapter 3 genes involved in neuronal protection such as erythropoietin and VEGF are activated rapidly at the transcriptional or post-transcriptional level through direct hypoxia or redox regulation pathways. Differences in ischemic vulnerability of neuronal subtypes correlate with the cellular decision to embark on apoptosis and the ultimate switch to programmed death may be influenced by a variety of modulators from intracellular signaling pathways. The exact molecular interactions that permit the mitogen activated protein kinase (MAPK) family of signaling molecules to modulate the progression to apoptosis are not yet known but the critical branches of these kinase cascades have been determined and it is certain for example that their components contribute to and modulate the regulation of the mitochondrial pathway of apoptosis. Signaling components of MAPK and/or apoptosis pathways have been characterized in species as diverse as worms, fruit flies, yeast, and man and are of central interest as potential mediators of the increased survival characteristics of anoxia tolerant species. Clinical implications of these pathways are becoming clear from experiments employing treatment with pharmacological antagonists and through transfer of therapeutic gene targeting vectors. A number of physiological mechanisms of neuronal protection such as adenosine

The Brain Without Oxygen. Third Edition. Peter L. Lutz, Göran E. Nilsson and Howard M. Prentice.

mediated tolerance and ATP dependent potassium channel activation are now known to be mediated through similar processes in the ischemic mammalian brain and the anoxia tolerant brain and a future challenge will be to link the molecular pro-survival and pro-death pathways with the physiological mechanisms.

2. IN VIVO MODELS OF CEREBRAL ISCHEMIA

Models of cerebral ischemia in mammals fall into two categories: global and focal ischemic insults. The four most widely used models of global ischemia are: 4 vessel occlusion in the rat, the two vessel occlusion combined with hypotension in the rat, two vessel occlusion in the gerbil and complete global ischemia or neck cuff in the rat or gerbil.

Almost all models of focal ischemia are based on occlusion of one middle cerebral artery. Models of permanent focal ischemia involve arterial blockage for the entire experiment, usually for one or more days.[18] In temporary focal occlusion models arteries are blocked for up to 3 h followed by restoration of the blood supply. Global ischemic episodes are generally kept short, usually between 5-20 min, and are associated with a rapid fall in ATP.[37] In the rat occlusion models of global ischemia ATP falls to less than 10% of normal values by 10 min. After short global ischemia levels of ATP can recover to at least 70% of normal[8], but tissue metabolic rate is depressed by at least 50% for 6 h indicating a prolonged functional impairment.[15] In global ischemia cell death is delayed, occurring between 12 h and several days after the insult.[37] In rat, 4 vessel occlusion for 10 min caused near total death of CA1 pyramidal cells in hippocampus within 3 days and a 5 min occlusion caused 50% cell death in the same area.[35] The gerbil hippocampus appears to be more sensitive to ischemic damage than that of the rat.[37]

3. MOLECULAR SIGNALING PATHWAYS REGULATING DEATH AND SURVIVAL IN CEREBRAL ISCHEMIA

A number of the molecular signaling pathways involved in ischemic and anoxic neuronal damage are understood and the relative involvement of distinct pathways for death depend on the severity and duration of the ischemic insult. The most immediate responses of neurons to severe insults such as anoxia or cell trauma generally result in necrosis, seen as cellular and organellar swelling followed by disruption of organelles, rupture of the plasma membrane and random DNA degradation.[5] However neurons that survive an acute ischemic episode may undergo delayed cell death which occurs through apoptosis.[39] Cellular apoptosis is characterized by shrinkage of the cytoplasm and nuclear condensation. This is followed by fragmentation of the nucleus and fusion of the endoplasmic reticulum with the plasma membrane which creates abnormal cell surface vesicles. Apoptosis culminates in cellular fragmentation resulting in the formation of membrane bound apoptotic bodies containing cytoplasmic organelles and nuclear fragments. Evidence of neuronal apoptosis in ischemia has been derived from a number of model systems. For example, in experimental transient global ischemia there is a loss of neurons through apoptosis like processes beginning 48-72 hours after the onset of the ischemic episode.[23] Using a short (30 minute) ischemic episode in the middle cerebral artery occlusion model of focal ischemia a large lesion was generated at 14 days that equaled in size that produced more rapidly by a severe necrotic occlusion.[14] The delayed neuronal death caused by this mild ischemia could be prevented by the translational inhibitor cyclohexamide indicating that protein synthesis was required to produce this neuronal loss.

4. GENE ACTIVATION AND TRANSCRIPTION FACTOR INDUCTION IN CEREBRAL ISCHEMIA

Within the first 30 min after global or of focal ischemia there is a rapid increase in expression of a number of immediate early genes (IEGs) including c-fos, c-jun, jun-B, krox-20 and zif-268, in a range of areas in the CNS, including both regions of direct injury and areas outside the initial injury.[2,31] After 4 vessel occlusion in the rat, for example, c-fos mRNA is induced in the dentate granule cells, the CA1 and CA3 pyramidal neurons in hippocampus, the neocortex, and the Purkinje cells of the cerebellum.[58] In models of global ischemia, protein synthesis is known to be decreased. In global ischemia therefore translation of mRNAs for IEGs may not be uniform. While the IEG proteins are detected in most areas, in CA1 neurons which are highly vulnerable to ischemia there was no detection of c-fos and zif268 proteins.[32] Successful translation of IEGs has been reported in models of focal cerebral ischemia. The prototype immediate early gene is c-fos which encodes a 380 amino acid protein, the Fos protein. This contains both a basic region that is responsible for DNA binding and a leucine zipper motif which is a dimerisation domain. The leucine zipper contains leucine residues every 7 amino acids creating a motif that enables the protein to align with other proteins containing this structure to form dimers.[2] Functional gene activation by c-fos and c-jun depends on dimerisation into the activator protein-1 (AP-1) complex. In focal ischemia, functional AP-1 binding to the promoter regions of responsive genes has been reported.[38] Although mRNA expression for IEGs has been found in both the core and the penumbral regions in focal ischemia, AP-1 promoter binding activity was shown to be localized specifically to the penumbra.[49] This result may reflect an inhibition of protein synthesis in the severely affected core of the focal ischemia.

5. ACTIVATION OF HEAT SHOCK PROTEIN (HSP) GENES

The induction of HSPs in focal or global ischemia is of special interest because they are specifically induced in cells responding to injury.[52] HSPs also mediate protective adaptations in ischemic tissues. For example HSPs function as molecular chaperones and can protect proteins from denaturation. HSPs are also capable of renaturation of denatured proteins.[57] The number after the abbreviated name of different HSPs refers to the molecular weight of the particular subtype of the protein. Hsp27, Hsp70, Hsp72 and Hsp90 are of particular significance in the brain where they represent markers of specific stress responses. Another heat shock protein Hsc70 was originally named on the basis of its constitutive pattern of expression but it is now known to be inducible in a number of tissues including brain.

The transcriptional activation of HSP genes is regulated by heat shock factors (HSFs) which bind to Hsp90 in non-stressed cells. Once proteins become denatured Hsp90 binds to the denatured proteins releasing the HSFs. The HSFs are subsequently activated by phosphorylation after which they bind to the promoter regions of the HSP genes.[52,60]

In global ischemia in the gerbil mRNAs for Hsp70, Hsc70 and Hsp90 rise rapidly in all regions and persist. HSP proteins are generally elevated over a few hours. However an interesting exception is seen with the vulnerable CA1 region where no induction is seen[30]. In marked contrast to the gerbil, the ischemic rat brain expresses Hsp70 at high levels in the CA1 region. In temporary focal ischemia, neurons in the ischemic core do not show Hsp70 immunoreactivity whereas neurons in the penumbra, many of which may recover, show expression of the protein. The neuronal expression of Hsp70 has been regarded as a molecularly defined penumbra of protein denaturation. Expression of HSPs tends to be localized in regions of cell survival and associated with brain areas of tissue protection against cell damage. The areas expressing Hsp70 protein correspond to zones where proteins are becoming denatured and where attempts are made at protein renaturation.[51] Hsp70 is also involved in protection of tissues against ischemic injury and has been implicated in ischemic preconditioning, a phenomenon where a short mild ischemic injury affords protection against a later severe injury, in both heart and hippocampus (see Chapter 9). It has recently been demonstrated that transgenic mice with overexpression of Hsp70 in brain are resistant to tissue damage from strokes by permanent middle cerebral artery occlusion again reinforcing the role of HSPs in tissue protection against ischemic damage.[48]

6. SIGNALING MOLECULES IN CEREBRAL ISCHEMIA

6.1 MAP Kinase Signaling

The members of the mitogen activated protein kinase (MAPK) superfamily of signaling molecules are highly conserved in evolution. These protein kinases preferentially phosphorylate serine/threonine residues within a motif that is flanked by prolines and is minimally a serine/threonine-proline motif. There are at least three principal pathways characterized by their family members: the extracellular responsive kinases (ERKs, also known as the p42/p44-MAPkinases), the c-jun N-terminal kinases which are also known as the stress-activated kinases (SAPK/JNKs), and the p38-MAPKs[54] (Fig. 5.1). The p38-MAPKs and SAPK/JNKs are activated by stress signals such as inflammatory cytokines, heat-shock and ischemia. By contrast the ERK pathway is activated in response to growth factors, oxidative stress, increases in intracellular calcium levels or glutamate receptor stimulation.[34,53]

MAPK is the final of 3 components in a phosphorylation cascade. MAPKs are activated by MAPK kinases (MAPKKs) which catalyze the phosphorylation of a tyrosine and a threonine residue on the MAPK.

The MAPKKs have in turn been phosphorylated on serine/threonine residues by MAPKK kinases (MAPKKKs).[54]

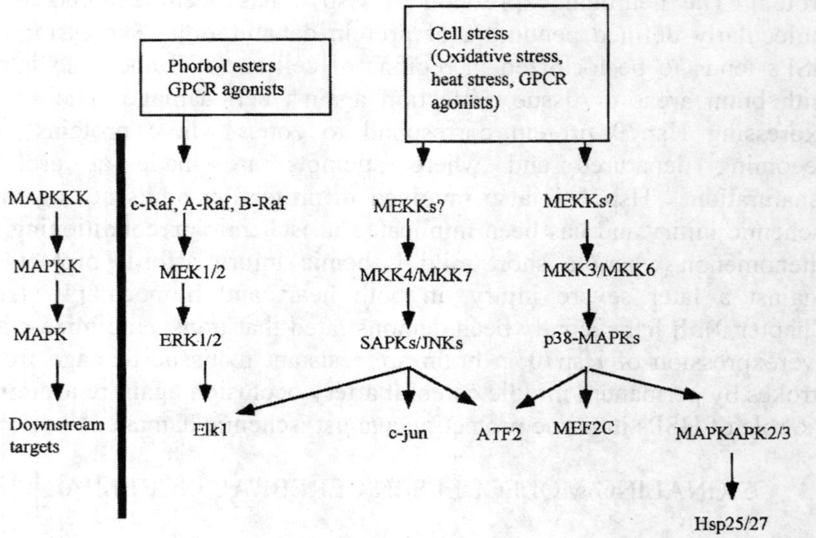

Fig. 5.1 The mitogen activated protein kinase pathways. The three main mitogen activated protein kinase pathways showing the different levels of regulation in these kinase cascades (GPCR: G-protein coupled receptor).

MEKs 1 and 2 are the MAPKKs for the ERK pathway. Recent studies on focal cerebral ischemia in mice suggest that the ERK pathway may play a role in inducing ischemic damage.[3] When mice were treated with a MEK1 inhibitor, focal infarct volume at 22 hours after ischemia was decreased by more than 50%. The ERK pathway was activated during the ischemia but ERK activity was reduced by administration of the inhibitor. By contrast other studies[19,21] point to a role for the ERK pathway in ischemia tolerance. ERKs may exert their influence on ischemia tolerance through activation of immediate early gene responses.

Generally, unlike activation of the ERK pathway which may mediate survival responses, activation of SAPK/JNK and p38 pathways promote cell death. In gerbil hippocampus exposed to transient forebrain ischemia inhibition of p38-MAPKs protected CA1 cells against delayed neuronal death.[55] The role of p38-MAPKs in cerebral ischemia remains controversial with distinct ischemic models resulting in a differential cell type pattern of p38 expression. In the rat focal cerebral ischemia model (transient or permanent) activated p38-MAPK was increased in astrocyte like cells in both the core and penumbra region.[28]

A range of studies have implicated c-jun N-terminal kinase (SAPK/JNK) activation in neuronal death following ischemia. A downstream target of the SAPK/JNK is the transcription factor c-jun which forms a dimerisation partner of c-fos in constituting the activator protein-1 (AP-1) complex. In the rat model of middle cerebral artery occlusion, phosphorylated c-jun expression was detected in neurons in the necrotic infarct areas of the striatum, piriform and entorhinal cortex at 3 h post-infarct with continued expression at 72 h post infarction.[27] SAPK/JNK activity was increased in hippocampus and cortex at 24 h post infarct. Expression of phosphorylated SAPK/JNK was demonstrated in another study at 8 h following middle cerebral artery occlusion.[25,24] In global forebrain ischemia SAPK/JNK is phosphorylated by 24 h after the ischemic insult.[17]

6.2 Cell Cycle Regulators

There is an increasing body of evidence that alterations in cell cycle regulatory pathways are involved in delayed neuronal death in cerebral ischemia.[29] Increases in expression of the cell cycle components cyclin D1 and cyclin dependent kinase 4 (Cdk4) have been demonstrated after focal ischemia in rats and mice.[36] Aberrant activation of cell cycle processes may cause apoptosis in post-mitotic neurons following an ischemic insult. The cell cycle is a highly coordinated process that depends on the activation of cyclin dependent kinases (CDKs). Regulation of CDKs depends on binding to the appropriate cyclin partner and, in combination, specific CDK/cyclin complexes are involved in controlling progressive phases of the cell cycle. One important downstream target of CDKs is the tumor suppressor protein, retinoblastoma protein (pRb), which upon hyperphosphorylation becomes released from the transcription factor complex, E2F1, allowing activation of genes required for S phase transition. One recent study on focal stroke in rats has demonstrated an increase in cyclin D1 levels and in phosphorylation of its target pRB.[47] A deregulated increase in E2F1 was also observed in this model. Interestingly administration of a CDK inhibitor blocks pRb phosphorylation and the elevation in E2F1 concomitant with an 80% reduction in neuronal death.

The re-emergence of molecular developmental events in the ischemic brain has also led to examination of the exciting possibility of an increase in neuronal number after stroke.[10] There have been recent demonstrations of neurogenesis in healthy adult rodent and primate brains. A post-ischemic increase in cell cycle proteins including the neuronal determination protein neuroD has been interpreted as possible evidence for activation of neuronal stem cells and progenitor cells from a quiescent state. In light of very recent reports of the existence of stem cells in a wide variety of adult tissues including the brain, the possibility that ischemia may elicit an increase in neural stem cell number points to important novel avenues for therapies directed at replacing the depleted neuronal population.

7. MEDIATORS AND MODULATORS OF APOPTOSIS

7.1 Excitatory Amino Acids and Apoptosis

It has been long understood that exposure of neurons to high concentrations of glutamate leads to rapid necrotic cell death through overactivation of NMDA receptors and Ca^{2+} influx (see Chapter 4). In recent studies however it has become clear that when acute effects of glutamate on neurons are blocked it is possible to examine a delayed effect of glutamate exposure on eliciting apoptotic cell death.[22,46,5] Thus the elevated extracellular glutamate observed in a variery of models of cerebral ischemia may elicit processes of apoptosis as well as necrotic cell death. Granular cell apoptosis has been observed in culture models after moderate glutamate exposure as well as after selective AMPA receptor stimulation.[33] Acute permanent vascular occlusion will generally result in necrosis whereas a milder injury will elicit apoptosis particularly in the ischemic penumbra.

7.2 Bcl-2 Family Members

The commitment to apoptosis is typically determined by the opposing influences of members of the Bcl-2 family. There are at least 15 Bcl-2 family members in mammalian cells all of which contain at least one of four conserved motifs known as a Bcl-2 homology domains (BH1 to 4). The pro- and anti-apoptotic Bcl-2 family members integrate upstream survival and stress signals to determine whether to initiate the switch to apoptosis.[1,20] Most pro-survival Bcl-2 family members contain at least domains BH1 and BH2. The pro-apoptotic family members fall into two classes: those that resemble Bcl-2 closely such as Bax which has BH1, BH2 and BH3, and another seven pro-apoptotic family members that possess only the BH3 domain[1]. Bcl-2 family member proteins are located in the membranes of the endoplasmic reticulum, the nuclear envelope and outer mitochondrial membrane. Following the "decision" to initiate the cell death pathway, critical down-stream target proteins are excised through the action of a class of proteases known as the caspases.

Apoptosis is characterized in a wide range of cellular systems by two pathways, the death receptor pathway, activated by cytokines of the TNF-alpha family and the mitochondrial pathway which is activated by cellular and environmental cues that are translated by sentinel proteins including Bim a Bcl-2 homologue.[1] A molecule that has been implicated in signaling between the death receptor pathway and the mitochondrial pathway is a pro-apoptotic Bcl-2 family member known as Bid (see Fig. 5.2). With regard to the death receptor pathway it has been shown in many studies that TNF-alpha a death receptor ligand is endogenously released during ischemia and is toxic in both transient and permanent ischemia.[43] One drug, CNI-1493, a recently described tetravalent guanylhydrazone compound that specifically inhibits cerebral TNF-alpha synthesis, resulted in a dramatic reduction of 75% in infarct volume, thus indicating an important causative role of TNF-alpha in post ischemic damage.[41]

The molecules that mediate apoptosis include death receptors, components of the mitochondrial pathway such as Bcl-2 and its antagonist Bax, as well as the executioners of programmed cell death, the caspases. Bcl-2 and its homologue Bcl-x(long) (Bcl-xl), another Bcl-2 family member, function as suppressors of apoptosis.[9] Bcl-2 may directly regulate the very early events in mitochondrial disturbances associated with apoptosis such as changes in pH, membrane permeability and outer membrane integrity.[20] This regulation will affect downstream pathways for proteins escaping from the intermembrane space such as cytochrome c which acts as a cofactor for the activation of caspase 9 by the adaptor or scaffold protein known as Apaf-1.[7]

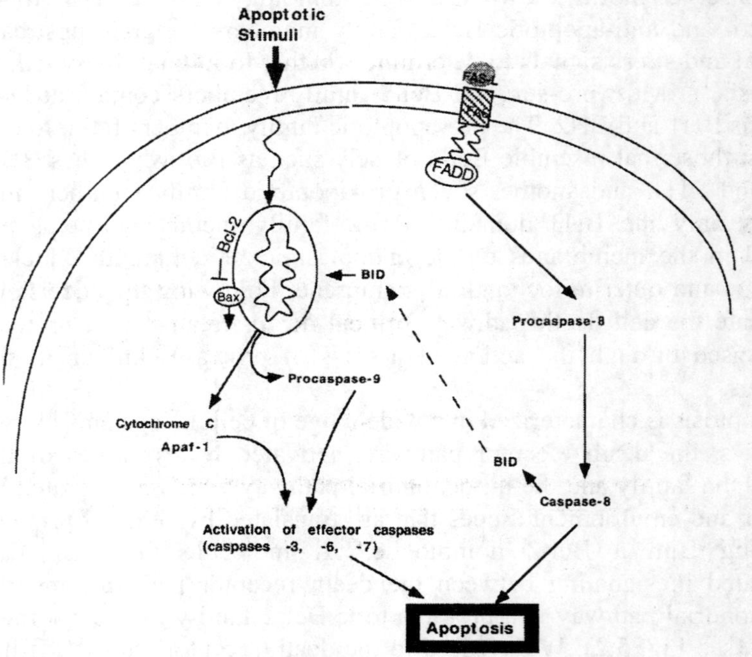

Fig. 5.2 The mitochondrial and death receptor pathways of apoptosis summarizing the principal components of each pathway.

The 3D structure of Bcl-xl has been found to resemble that of membrane-penetrating bacterial toxins which has pointed to the possibility that Bcl-2 family members may form pores in organelle membranes.[42] As an alternative to forming pores themselves, Bcl-2 family members may stabilize or modify an exisiting channel, such as the permeability transition pore through which adenine nucleotides and other small molecules can traverse the mitochondrial membrane. This channel is composed of key components including the voltage gated anion channel (VDAC) in the outer mitochondrial membrane, the adenine nucleotide translocator in the inner mitochondrial membrane and cyclophilin D in the mitochondrial matrix.[11] Bcl-xl and Bax may be able to associate with VDAC whereby Bcl-xl might close the channel and Bax would keep it open to allow cytochrome c to pass through and subsequently participate in activation of caspase 9 by the adaptor protein Apaf-1.[56]

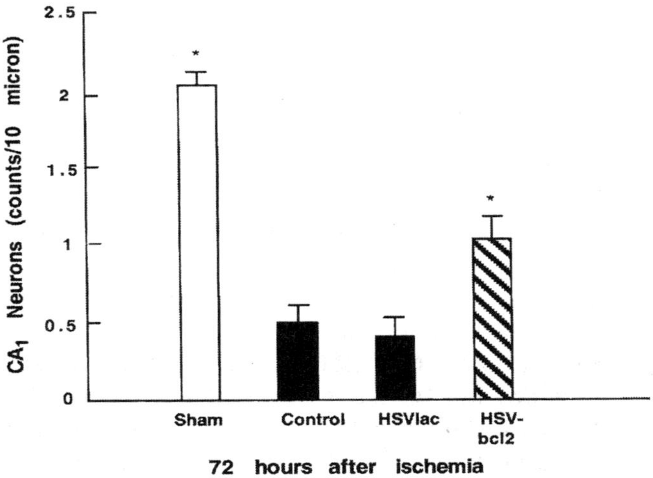

*Fig. 5.3 Neuronal protection by Bcl-2. Number of CA1 neurons 72 hours after ischemia. Infusion of a viral expression vector (Herpes simplex virus) encoding Bcl-2 significantly protected CA1 neurons at 72 h after transient global ischemia by comparison to control ischemia and ischemic controls transduced with a beta-galactosidase (HSV-lac) vector (*p<0.05 differs from control) (after Antonawich et al., 1999).*

In brain the Bcl-2 family of proteins follow a similar pattern to HSP in terms of upregulation in non-vulnerable regions of hippocampus after global ischemia. Indeed Bcl-2 represents an important candidate for a protectant molecule against ischemic damage in both rat and gerbil models. Overexpression of Bcl-2 through infusion of a viral expression vector into the CA1 region of the gerbil hippocampus led to enhanced survival of neurons in this region 72 hours after transient global ischemia.[4] In rat subjected to 15 minutes of global cerebral ischemia[9] expression of Bcl-2 and Bcl-xl (the predominant form of Bcl-x in brain) were found to be induced in CA3 and dentate cells but not in vulnerable CA1 hippocampal cells. Two of the homologues of Bcl-2, known as Bax and Bcl-x(short) (Bcl-xs) are known to have a pro-apoptotic role. Bcl-xs results from the alternative splicing of the Bcl-x gene transcripts into a long form and a short form which lacks the transmembrane domain. Bcl-xs acts as an antagonist of both Bcl-2 and Bcl-xl. In global ischemia models in rat and gerbil the two pro-apoptotic proteins Bax and Bcl-xs were induced in CA1 neurons in hippocampus.[9,13] Bax was not increased in CA3 but both Bax and Bcl-xs were enhanced in a less vulnerable brain area, the dentate gyrus.[16]

7.3 *Caspases and Inhibitor of Apoptosis Proteins (IAPs)*

The caspase family of cysteine proteases play a key role as effectors of programmed cell death. Caspases can be subdivided into those involved in regulating other caspases (caspase 1, 4, 5, 8 and 9) and caspases responsible for downstream cellular disassembly, (caspases 3, 7 and 14) or both (caspases 2 and 10).[45] Caspase enzymes demonstrate a specificity for aspartic acid on the N-terminal side of the substrate cleavage site. The enzymes are characterized by having a specific pentapeptide in the catalytic site.[50] Caspases in their inactive state exist as proenzymes with an N-terminal prodomain together with one large and one small subunit. Functional caspase enzymes are activated by proteolytic processing. Caspase substrates include DNA repair enzymes, cell cycle regulatory and structural proteins and members of the Bcl-2 protein family. While effector caspases are activated by modification by upstream caspases the apical or initiator caspase is self-catalyzed. Caspase 9, an initiator caspase, is autocatalyzed through a multimerisation process aided by the adaptor protein Apaf-1.[26]

Recent experiments have demonstrated that caspase processing plays a critical role in cell death in models of cerebral ischemia. In non-ischemic animals caspases 1, 2, and 3 are constitutively expressed in the cortex and cubcortical regions. After transient global ischemia, in situ hybridization data has demonstrated an upregulation of caspase-3 mRNA in CA1 pyramidal neurons at 24 h, which persisted for at least 24 h. Activation of caspase-3 protein is detected in neurons in global ischemia suggesting a direct involvement of this protein in ischemic damage.[44] Caspase-1 is induced in models of global cerebral ischemia but its expression is only observed in glial cells. It is likely that the involvement of caspase-1 in ischemic damage is indirect and the protein may mediate inflammatory responses under conditions of global cerebral ischemia.[6]

The inhibitor of apoptosis proteins (IAPs) have been shown to participate in tolerance to cerebral ischemia through direct action on downstream caspases. One member of the IAP family, neuronal apoptosis inhibitory protein (NIAP) becomes elevated in neurons resistant to transient forebrain ischemia. A second family member, the X-chromosomal IAP (XIAP) causes direct inhibition of caspase 3 and caspase 7.[59]

8. SUMMARY

To conclude, ischemia in the brain, whether severe or mild, may result in induction of multiple cell signaling pathways and gene transcriptional events whose impact can involve cellular adaptations for survival or alternatively catastrophic responses including necrosis and apoptosis. Both mitochondrial and death receptor pathways play a significant role in neuronal apoptosis in ischemic conditions and the extent to which these pathways are activated will vary depending on the brain region affected, with specific cellular populations in the hippocampus for example displaying different vulnerabilities to hypoxia/ischemia mediated damage. Specific families of signaling molecules are capable of conferring cellular protection against ischemic damage and these include the heat shock proteins which are induced in models of focal and global ischemia and which mediate resistance to stroke induced tissue damage in transgenic mouse models. The various MAP kinase pathways are activated in cerebral ischemia and may influence tissue damage, the p38-MAPK pathway generally by mediating survival responses and the ERK and SAPK/JNK pathways in contributing to ischemic cell death in part through apoptosis signaling. Recent evidence of regulation of cell cycle components in neurons following ischemia raises important considerations regarding the possibility of neuronal regeneration or partial replenishment of neuronal populations following ischemia induced tissue damage. As intense research investigation progresses in the area of molecular adaptations in cerebral ischemia our increased understanding of the mechanisms of neuronal cell signaling is pointing towards components of stress-kinase, apoptosis and cell cycle pathways as some of the most promising molecular targets for future therapeutic interventions.

9. REFERENCES

1. Adams JM, Cory S. Life-or-death decisions by the Bcl-2 protein family. Trends Biochem Sci. 2001 Jan;26(1):61-6
2. Akins PT, Liu PK, Hsu CY. Immediate early gene expression in response to cerebral ischemia. Friend or foe? Stroke. 1996 Sep;27(9):1682-7.
3. Alessandrini A, Namura S, Moskowitz MA, Bonventre JV. MEK1 protein kinase inhibition protects against damage resulting from focal cerebral ischemia.Proc Natl Acad Sci U S A. 1999 Oct 26;96(22):12866-9.
4. Antonawich FJ, Federoff HJ, Davis JN. BCL-2 transduction, using a herpes simplex virus amplicon, protects hippocampal neurons from transient global ischemia. Exp Neurol. 1999 Mar;156(1):130-7.
5. Banasiak KJ, Xia Y, Haddad GG. Mechanisms underlying hypoxia-induced neuronal apoptosis. Prog Neurobiol. 2000 Oct;62(3):215-49.
6. Bhat RV, DiRocco R, Marcy VR, Flood DG, Zhu Y, Dobrzanski P, Siman R, Scott R, Contreras PC, Miller M. Increased expression of IL-1beta converting enzyme in hippocampus after ischemia: selective localization in microglia. J Neurosci. 1996 Jul 1;16(13):4146-54.
7. Budihardjo I, Oliver H, Lutter M, Luo X, Wang X. Biochemical pathways of caspase activation during apoptosis. Annu Rev Cell Dev Biol. 1999;15:269-90.
8. Cardell M, Koide T, Wieloch T. Pyruvate dehydrogenase activity in the rat cerebral cortex following cerebral ischemia. J Cereb Blood Flow Metab. 1989 Jun;9(3):350-7.
9. Chen J, Graham SH, Nakayama M, Zhu RL, Jin K, Stetler RA, Simon RP. Apoptosis repressor genes Bcl-2 and Bcl-x-long are expressed in the rat brain following global ischemia. J Cereb Blood Flow Metab. 1997 Jan;17(1):2-10.
10. Cramer SC, Chopp M. Recovery recapitulates ontogeny. Trends Neurosci. 2000 Jun;23(6):265-71.
11. Crompton M. Mitochondrial intermembrane junctional complexes and their role in cell death. J Physiol. 2000 Nov 15;529 Pt 1:11-21.
12. Dirnagl U, Iadecola C, Moskowitz MA. Pathobiology of ischaemic stroke: an integrated view. Trends Neurosci. 1999 Sep;22(9):391-7.
13. Dixon EP, Stephenson DT, Clemens JA, Little SP. Bcl-Xshort is elevated following severe global ischemia in rat brains. Brain Res. 1997 Nov 21;776(1-2):222-9.
14. Du C, Hu R, Csernansky CA, Hsu CY, Choi DW. Very delayed infarction after mild focal cerebral ischemia: a role for apoptosis? J Cereb Blood Flow Metab. 1996 Mar;16(2):195-201.
15. Edmonds HL Jr, Wauquier A, Melis W, Van den Broeck WA, Van Loon J, Janssen PA. Improved short-term neurological recovery with flunarizine in a canine model of cardiac arrest. Am J Emerg Med. 1985 Mar;3(2):150-5.
16. Gillardon F, Lenz C, Waschke KF, Krajewski S, Reed JC, Zimmermann M, Kuschinsky W. Altered expression of Bcl-2, Bcl-X, Bax, and c-Fos colocalizes with DNA fragmentation and ischemic cell damage following middle cerebral artery occlusion in rats. Brain Res Mol Brain Res. 1996 Sep 1;40(2):254-60.
17. Gillardon F, Spranger M, Tiesler C, Hossmann KA. Expression of cell death-associated phospho-c-Jun and p53-activated gene 608 in hippocampal CA1 neurons following global ischemia. Brain Res Mol Brain Res. 1999 Nov 10;73(1-2):138-43.
18. Ginsberg MD, Busto R. Rodent models of cerebral ischemia. Stroke. 1989 Dec;20(12):1627-42.
19. Gonzalez-Zulueta M, Feldman AB, Klesse LJ, Kalb RG, Dillman JF, Parada LF, Dawson TM, Dawson VL. Requirement for nitric oxide activation of p21(ras)/extracellular regulated kinase in neuronal ischemic preconditioning. Proc Natl Acad Sci U S A. 2000 Jan 4;97(1):436-41.
20. Green DR, Reed JC. Mitochondria and apoptosis. Science. 1998 Aug 28;281(5381):1309-12.

21. Gu Z, Jiang Q, Zhang G, Cui Z, Zhu Z. Diphosphorylation of extracellular signal-regulated kinases and c-Jun N-terminal protein kinases in brain ischemic tolerance in rat. Brain Res. 2000 Mar 31;860(1-2):157-60.
22. Gwag BJ, Lobner D, Koh JY, Wie MB, Choi DW. Blockade of glutamate receptors unmasks neuronal apoptosis after oxygen-glucose deprivation in vitro. Neuroscience. 1995 Oct;68(3):615-9.
23. Halterman MW, Federoff HJ. HIF-1alpha and p53 promote hypoxia-induced delayed neuronal death in models of CNS ischemia. Exp Neurol. 1999 Sep;159(1):65-72.
24. Harper SJ, LoGrasso P. Signalling for survival and death in neurones: the role of stress-activated kinases, JNK and p38. Cell Signal. 2001 May;13(5):299-310.
25. Hayashi T, Sakai K, Sasaki C, Zhang WR, Warita H, Abe K. c-Jun N-terminal kinase (JNK) and JNK interacting protein response in rat brain after transient middle cerebral artery occlusion. Neurosci Lett. 2000 Apr 28;284(3):195-9.
26. Hengartner MO. The biochemistry of apoptosis. Nature. 2000 Oct 12;407(6805):770-6.
27. Herdegen T, Claret FX, Kallunki T, Martin-Villalba A, Winter C, Hunter T, Karin M. Lasting N-terminal phosphorylation of c-Jun and activation of c-Jun N-terminal kinases after neuronal injury. J Neurosci. 1998 Jul 15;18(14):5124-35.
28. Irving EA, Barone FC, Reith AD, Hadingham SJ, Parsons AA. Differential activation of MAPK/ERK and p38/SAPK in neurones and glia following focal cerebral ischaemia in the rat. Brain Res Mol Brain Res. 2000 Apr 14;77(1):65-75.
29. Katchanov J, Harms C, Gertz K, Hauck L, Waeber C, Hirt L, Priller J, von Harsdorf R, Bruck W, Hortnagl H, Dirnagl U, Bhide PG, Endres M. Mild cerebral ischemia induces loss of cyclin-dependent kinase inhibitors and activation of cell cycle machinery before delayed neuronal cell death. J Neurosci. 2001 Jul 15;21(14):5045-53.
30. Kawagoe J, Abe K, Sato S, Nagano I, Nakamura S, Kogure K. Distributions of heat shock protein-70 mRNAs and heat shock cognate protein-70 mRNAs after transient global ischemia in gerbil brain. J Cereb Blood Flow Metab. 1992 Sep;12(5):794-801.
31. Kiessling M, Gass P. Stimulus-transcription coupling in focal cerebral ischemia. Brain Pathol. 1994 Jan;4(1):77-83.
32. Kiessling M, Stumm G, Xie Y, Herdegen T, Aguzzi A, Bravo R, Gass P. Differential transcription and translation of immediate early genes in the gerbil hippocampus after transient global ischemia. J Cereb Blood Flow Metab. 1993 Nov;13(6):914-24.
33. Larm JA, Cheung NS, Beart PM. Apoptosis induced via AMPA-selective glutamate receptors in cultured murine cortical neurons. J Neurochem. 1997 Aug;69(2):617-22.
34. Lewis TS, Shapiro PS, Ahn NG. Signal transduction through MAP kinase cascades. Adv Cancer Res. 1998;74:49-139.
35. Li H, Buchan AM. Treatment with an AMPA antagonist 12 hours following severe normothermic forebrain ischemia prevents CA1 neuronal injury. J Cereb Blood Flow Metab. 1993 Nov;13(6):933-9.
36. Li Y, Chopp M, Powers C, Jiang N. Immunoreactivity of cyclin D1/cdk4 in neurons and oligodendrocytes after focal cerebral ischemia in rat. J Cereb Blood Flow Metab. 1997 Aug;17(8):846-56.
37. Lipton P. Ischemic cell death in brain neurons. Physiol Rev. 1999 Oct;79(4):1431-568.
38. Liu PK, Salminen A, He YY, Jiang MH, Xue JJ, Liu JS, Hsu CY Suppression of ischemia-induced fos expression and AP-1 activity by an antisense oligodeoxynucleotide to c-fos mRNA. Ann Neurol. 1994 Oct;36(4):566-76.
39. MacManus JP, Hill IE, Huang ZG, Rasquinha I, Xue D, Buchan AM.DNA damage consistent with apoptosis in transient focal ischaemic neocortex. Neuroreport. 1994 Jan 12;5(4):493-6.
40. MacManus JP, Linnik MD. Gene expression induced by cerebral ischemia: an apoptotic perspective. J Cereb Blood Flow Metab. 1997 Aug;17(8):815-32.
41. Meistrell ME 3rd, Botchkina GI, Wang H, Di Santo E, Cockroft KM, Bloom O, Vishnubhakat JM, Ghezzi P, Tracey KJ. Tumor necrosis factor is a brain damaging cytokine in cerebral ischemia. Shock. 1997 Nov;8(5):341-8.

42. Muchmore SW, Sattler M, Liang H, Meadows RP, Harlan JE, Yoon HS, Nettesheim D, Chang BS, Thompson CB, Wong SL, Ng SL, Fesik SW. X-ray and NMR structure of human Bcl-xL, an inhibitor of programmed cell death. Nature. 1996 May 23;381(6580):335-41.
43. Nawashiro H, Martin D, Hallenbeck JM. Neuroprotective effects of TNF binding protein in focal cerebral ischemia. Brain Res. 1997 Dec 19;778(2):265-71.
44. Ni B, Wu X, Su Y, Stephenson D, Smalstig EB, Clemens J, Paul SM. Transient global forebrain ischemia induces a prolonged expression of the caspase-3 mRNA in rat hippocampal CA1 pyramidal neurons. J Cereb Blood Flow Metab. 1998 Mar;18(3):248-56.
45. Nicholson DW, Thornberry NA. Caspases: killer proteases. Trends Biochem Sci. 1997 Aug;22(8):299-306.
46. Nicotera P, Leist M, Manzo L. Neuronal cell death: a demise with different shapes. Trends Pharmacol Sci. 1999 Feb;20(2):46-51.
47. Osuga H, Osuga S, Wang F, Fetni R, Hogan MJ, Slack RS, Hakim AM, Ikeda JE, Park DS. Cyclin-dependent kinases as a therapeutic target for stroke. Proc Natl Acad Sci U S A. 2000 Aug 29;97(18):10254-9.
48. Rajdev S, Hara K, Kokubo Y, Mestril R, Dillmann W, Weinstein PR, Sharp FR. Mice overexpressing rat heat shock protein 70 are protected against cerebral infarction. Ann Neurol. 2000 Jun;47(6):782-91.
49. Salminen A, Liu PK, Hsu CY. Alteration of transcription factor binding activities in the ischemic rat brain. Biochem Biophys Res Commun. 1995 Jul 26;212(3):939-44.
50. Schulz JB, Weller M, Moskowitz MA. Caspases as treatment targets in stroke and neurodegenerative diseases. Ann Neurol. 1999 Apr;45(4):421-9.
51. Sharp FR, Lu A, Tang Y, Millhorn DE. Multiple molecular penumbras after focal cerebral ischemia. J Cereb Blood Flow Metab. 2000 Jul;20(7):1011-32.
52. Sharp FR, Massa SM, Swanson RA. Heat-shock protein protection. Trends Neurosci. 1999 Mar;22(3):97-9.
53. Skaper SD, Facci L, Strijbos PJ. Neuronal protein kinase signaling cascades and excitotoxic cell death. Ann N Y Acad Sci. 2001 Jun;939:11-22.
54. Sugden PH, Clerk A.Cellular mechanisms of cardiac hypertrophy. J Mol Med. 1998 Oct;76(11):725-46.
55. Sugino T, Nozaki K, Takagi Y, Hattori I, Hashimoto N, Moriguchi T, Nishida E. Activation of mitogen-activated protein kinases after transient forebrain ischemia in gerbil hippocampus. J Neurosci. 2000 Jun 15;20(12):4506-14.
56. Vander Heiden MG, Chandel NS, Schumacker PT, Thompson CB. Bcl-xL prevents cell death following growth factor withdrawal by facilitating mitochondrial ATP/ADP exchange. Mol Cell. 1999 Feb;3(2):159-67.
57. Welch WJ.How cells respond to stress. Sci Am. 1993 May;268(5):56-64.
58. Wessel TC, Joh TH, Volpe BT. In situ hybridization analysis of c-fos and c-jun expression in the rat brain following transient forebrain ischemia. Brain Res. 1991 Dec 20;567(2):231-40.
59. Xu DG, Crocker SJ, Doucet JP, St-Jean M, Tamai K, Hakim AM, Ikeda JE, Liston P, Thompson CS, Korneluk RG, MacKenzie A, Robertson GS. Elevation of neuronal expression of NAIP reduces ischemic damage in the rat hippocampus. Nat Med. 1997 Sep;3(9):997-1004.
60. Zou J, Guo Y, Guettouche T, Smith DF, Voellmy R. Repression of heat shock transcription factor HSF1 activation by HSP90 (HSP90 complex) that forms a stress-sensitive complex with HSF1. Cell. 1998 Aug 21;94(4):471-80.

CHAPTER 6

BRAINS THAT SURVIVE: ADAPTATIONS TO ANOXIA

1. INTRODUCTION

Not all vertebrates die rapidly in anoxia. There are some, albeit very few, exceptions to this rule. The two members of the cyprinid fish genus *Carassius,* the crucian carp *(C. carassius)* and the goldfish (*C. auratus*), and some North American freshwater turtles of the genera *Chrysemys* and *Trachemys* (formerly *Pseudemys*) have the extraordinary ability to survive without oxygen for weeks or even months at temperatures below 10 °C, and for days at 20-25 °C.[1,13,18,28] Fig. 6.1 clearly reveals that if anoxia survival time is plotted against body temperature the truly anoxia tolerant species form a well separated regression line compared to other vertebrates. Indeed, the difference in survival time between the two groups is in the order of two magnitudes. Certain species of frog might be regarded as intermediate being able to survive brain anoxia for several hours, but, their neuronal accommodation to anoxia is so peculiar that it is treated as a "special case" in chapter 8.

However, even for the mammal, anoxia tolerance can be considerably increased under certain conditions. The mammalian neonate can withstand anoxia to a much greater degree than the adult. Enhanced hypoxia tolerance is seen during hibernation, and it is undoubtedly a factor in the survival of children from cold water submersion. During prolonged dives marine mammals can become severely hypoxic and some adaptations to hypoxia are shown by animals living at high altitude. Whether these "special cases" have mechanisms in common with the anoxia tolerant vertebrates will be treated separately in a subsequent chapter.

The Brain Without Oxygen. Third Edition. Peter L. Lutz, Göran E. Nilsson and Howard M. Prentice.

Anoxia-tolerant vertebrates are of special interest in the present context since experimental studies on these animals can show us how evolution has solved the problem of living without oxygen. Indeed, there are numerous questions that only can be answered by studying such organisms. In particular, anoxia-tolerant vertebrates can serve as models that survive instead of die in anoxia. Thus, we can make experimental manipulations of hypothesized survival mechanisms in order to reveal causal connections between cellular and physiological perturbations and neuronal death. Such experiments cannot be made in mammals, since there the "control situation" in anoxia is death.

If we find that the crucian carp and the freshwater turtle brains possess unique biochemical mechanisms for anoxic survival that makes them fundamentally different from other vertebrates, then such findings will probably be difficult to implement in a clinical situation, notwithstanding the intellectual satisfaction achieved in their discovery. But in fact, when the experimental results from studies of anoxia-tolerant vertebrates are compared to mammalian studies, the differences seem to be mainly quantitative, and basically function to avoid crossing death causing thresholds such as anoxic depolarization. The examples of truly qualitative differences between anoxia-tolerant and anoxia-intolerant vertebrates appear to be very few, and are probably not by themselves prerequisites for anoxic survival, since they are not common to both groups of anoxia-tolerant vertebrates. This could be expected from the considerable phylogenetic distance between the fish and the turtle, with thousands of anoxia-intolerant species between them and their common ancestor. For example, the crucian carp's ability to produce ethanol during anoxia is not shared by the freshwater turtles.

2. PHYLOGENETIC FACTORS

The exceptional anoxia tolerance of the crucian carp and the freshwater turtle brains cannot be attributed to the lesser metabolic requirements of reptiles and fishes compared to those of mammals. The differences in brain metabolic rate between anoxia-tolerant vertebrates and "normal" vertebrates are trivial (Fig 2.12 in chapter 2) when compared to the 1000 fold difference seen in anoxic survival time between these two groups (Fig. 6.1). Indeed, it should be pointed out that anoxia surviving brains are probably uncommon even in the "cold blooded" vertebrates. The brains of lizards (*Anolis sagrei*) are much less anoxic tolerant than those of the turtle *T.scripta*[17] and, when corrected for temperature, the rainbow trout brain shows virtually the same susceptibility to anoxic failure as the mammalian brain, with a similar time course for the onset of anoxic depolarization (chapter 4). Undoubtedly there are a few other anoxia species yet to be discovered among the ectotherms. The Californian blind goby (*Typhologobius californiensis*) actually appears to be close to the crucian carp and the freshwater turtles in its degree of anoxia- tolerance, having been found to survive 80 h of anoxia at 15 °C.[4] Unfortunately, the blind goby is a tiny animal, weighing less than 5 g, making detailed neurophysiological studies difficult. Some ectothermic species show an intermediate degree of anoxia tolerance. These include certain frogs and toads a few of the intertidal fishes and eels, tilapia and hagfish.[29,33] The European eel (*Anguilla anguilla*), for example, survives some 6 h of anoxia at 15 °C[31] and a tilapia species (*Sarotherodon mossambicus*) has been found to tolerate at least 2 h of anoxia at 20 °C.[30] The mechanisms that allow frogs to survive a few h of anoxia appears to be quite different from those of turtles and fish, and are therefore given special attention in Chapter 8.

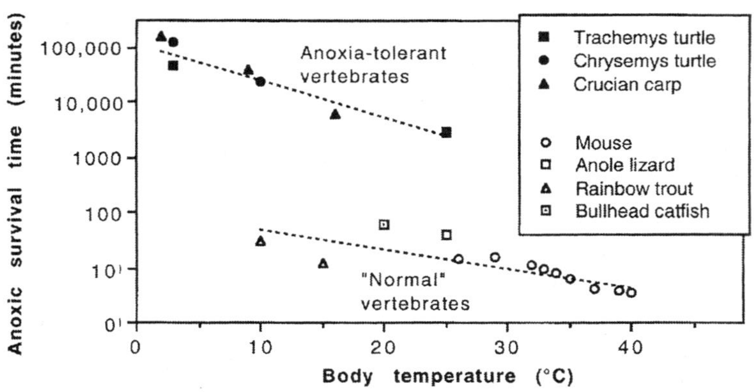

Fig. 6.1 Anoxic survival time of vertebrates at different temperatures. Two discrete groups are seen, one showing a more than 1000 fold higher capacity to survive anoxia than the other.

Anoxia tolerance is not, therefore, a simple consequence of a less complex level of organization in the brains of "lower" vertebrates, but is the result of physiological adaptations in those few species that are able to survive prolonged anoxia.

3. TRANSITION TO THE ANOXIC STATE

Before considering the true challenge of surviving prolonged anoxia, it is relevant to discuss some of the processes involved in dealing with a fall in O_2 availability during the transition to the anoxic state.

In contrast to mammals, turtles breath intermittently with variable periods of apnea. In consequence, turtle blood O_2 tension can vary greatly even under aerobic conditions, and arterial O_2 tensions (P_aO_2) have been recorded falling from 100 torr to 20 torr during routine dives[2], which for a mammal would signify fairly serious hypoxia. Since most routine dives are fully aerobic and because large changes in arterial blood O_2 are a normal occurrence in these animals, it would be expected that turtles possess highly sensitive mechanisms to regulate tissue energy supplies in the face of fluctuating oxygen tensions. During routine dives while the P_aO_2 falls, energy requirements are not likely to be decreased since the turtle remains alert and active and performs such tasks such as feeding and hunting that require integrated central nervous processing. If CNS activity is undiminished during diving then it is likely that brain ATP consumption is not reduced. This means that the turtle brain must be endowed with mechanisms to stabilize ATP levels throughout a wide range of P_aO_2 values.

Changes in the brain's oxidative state during hypoxia and the transition to anoxia have been studied by comparing the effect of changing brain tissue oxygen tension (P_tO_2) on the redox status of cytochrome aa_3 (the terminal oxygen binding enzyme complex in the respiratory chain) and brain metabolite concentrations in the turtle and the rat.[15,26] At a normoxic P_aO_2 for the rat (approximately 110 torr) and at a similar P_aO_2 for the turtle blood, cytochrome aa_3 was found to be more reduced in the turtle brain despite the fact that the P_tO_2 values were similar in both species (10-12 torr) (Fig. 6.2). There were also interesting differences between the rat brain and turtle brain in the relationship between P_tO_2 and the redox status cytochrome aa_3. In the rat comparatively little change in the redox state of cytochrome aa_3 is seen over a range of oxygen tension changes around the normoxic P_tO_2 and above, however when the P_tO_2 reaches about 80% of normoxic the redox state of cytochrome aa_3 falls precipitously. This indicates that if the mammal normally maintains P_tO_2 over a narrow range then little change in the cytochrome aa_3 redox status will occur. In the turtle brain, by contrast, cytochrome aa_3 shows a more linear relationship over a wider range of P_tO_2 such that any change in P_tO_2 will cause a corresponding the mammal a shift towards cytochrome aa_3 reduction may function as a warning of threatening hypoxic failure[34] and presumably serves to initiate compensatory responses. In the turtle brain, on the other hand, the cytochrome aa_3 redox state will track P_tO_2 changes over a normal breathing cycle (Fig. 6.2). Also, in the turtle brain, the respiratory chain is significantly more reduced at all oxygen tensions compared to the mammal. This may be related to the comparatively higher rate of glycolysis seen in turtle brain slices even under fully aerobic conditions.[19] If ATP consumption in the turtle brain is maintained shift in the cytochrome aa_3 redox status. It has been suggested that in during voluntary dives, then any decline in aerobic ATP production brought about by decreases in P_tO_2 must be offset by increased ATP production through anaerobic glycolysis and/or utilization of creatine phosphate stores. There is evidence that such a compensatory system does indeed operate in the turtle. The turtle brain shows large increases in lactate concentration (revealing enhanced anaerobic glycolysis) and decreases in the creatine phosphate level well before cytochrome aa_3 becomes fully reduced[14] and an early activation of glycolysis is also indicated by increases in hexoses and triose phosphate intermediates within the first hour of anoxia.[11] During this period the heat production of the isolated cerebellum from this species remains constant[20], indicating that glycolysis is progressively enhanced as the ability to produce ATP by oxidative phosphorylation falls in response to declining P_tO_2.

If a reduction in cytochrome aa_3 is indeed an index of O_2 adequacy, and is associated with the initiation of creatine phosphate depletion[24] or anaerobic glycolysis[9], then the changing redox state of turtle brain cytochrome aa_3 that would occur during a routine dive may be a mechanism to couple the glycolytic rate with mitochondrial respiration in a regulated manner in order to maintain ATP supply. The rat, by contrast, employs a strategy which ensures that brain P_tO_2 does not decline. As a consequence, the redox state of rat brain cytochrome aa_3 does not normally vary, unless the tissue becomes hypoxic whereupon there is a precipitous fall in the cytochrome aa_3 redox level and a rapid, emergency, and perhaps maximal, activation of anaerobic glycolysis.

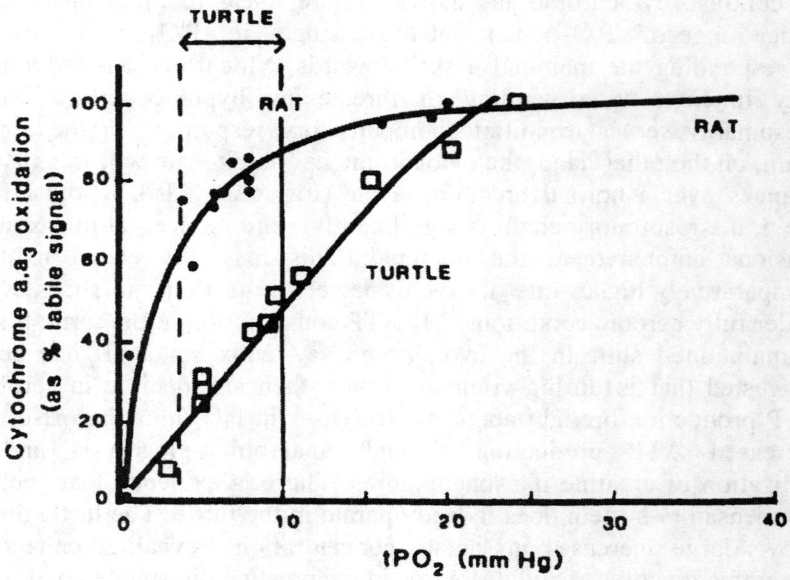

Fig. 6.2 Comparison of the relationships between brain oxygen tension (tPO_2) and cytochrome a,a_3 relationships in the turtle and rat. The solid vertical line indicates normoxic brain oxygen tension for the turtle and rat. The broken vertical line indicates the expected decrease in turtle brain oxygen tension that would occur during a dive. Adapted from Sick et al.[26] and Lutz et al.[15]

4. THE ANOXIC STATE

Before attempting to understand the mechanisms for anoxia tolerance we first must establish the basic parameters of the anoxic condition. Which facets of brain physiology are maintained, which change and to what degree do they change? This information will not only identify the key features which must be conserved for survival but will indicate those that can be compromised and allow us to construct hypotheses on the processes involved.

4.1 Ionic Gradients

One of the most striking differences between anoxia-tolerant and anoxia-intolerant animals is the ability of the latter to maintain brain ionic gradients. As discussed in the previous chapter, within a few minutes of anoxia the mammalian brain suffers an abrupt loss of transmembrane ion gradients and the resultant depolarization is a major cause of cell death. Basically anoxia-tolerant animals simply avoid depolarization. For example, in the turtle brain extracellular K^+ rises only by 1-2mM during 4 h anoxia and reached 3-6 mM during 48hr anoxia[15] (Fig. 6.3A). A similar capacity to maintain brain ionic gradients during prolonged anoxia is seen in the anoxic tolerant crucian carp[16] (Fig. 6.3B). In the rat brain, by contrast, ionic gradients collapse within 2 min of anoxia (Fig. 6.3A), and the brain of the rainbow trout is similarly vulnerable (Fig. 6.3B).

4.2 Energy Status

Equally striking is the ability of the anoxic tolerant brain to maintain its energy status. While ATP is depleted in the mammalian brain within seconds of anoxia (see chapter 2), ATP levels are held in the *in situ* turtle brain for at least 3-5 hr anoxia[14,11], or for 6 h anoxia in the isolated turtle cerebellum.[21] Similarly, the goldfish maintains brain ATP levels for 12 h of anoxia at 20 °C[29], and crucian carp brain slices maintain their ATP levels for at least 20 h of anoxia at 12 °C[10] (all the time frames given being dependent on the endurance of the experimenters rather than that of the animal). Like the mammal however, creatine phosphate is reduced during anoxia albeit at a much slower rate, reaching in the turtle brain about 50% of control values after 20 min anoxia.[12]

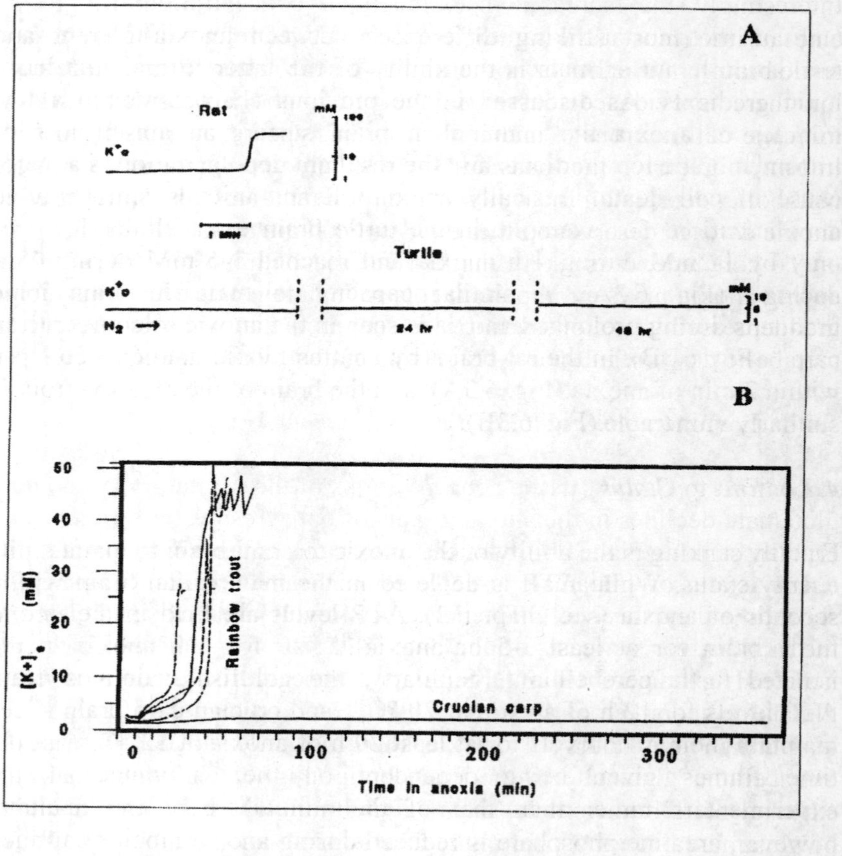

Fig. 6.3A Changes in brain extracellular K^+ during anoxia in the turtle (telencephalon) and rat (cerebral cortex). From Sick et al.[26]
6.3B. Changes in extracellular K^+ during anoxia in crucian carp and rainbow trout brain (optic lobes). From Nilsson et al.[18]

4.3 Electrical Activity

While in the mammal EEG and evoked potential activity becomes fully suppressed within seconds of anoxia, integrated electrical activity appears to continue in the anoxic turtle brain although at lesser levels. When the freshwater turtle brain becomes anoxic, the EEG has been shown to lose much of its normal high amplitude component within a few minutes[7,25], and evoked potential becomes reduced[25]. Similarly in the isolated turtle cerebellum, evoked field potentials are decreased by 80% on 4 h of anoxia.[21] In both the slice and the in situ preparation electrical activity fully recovers after reoxygenation[26,34] However, action potentials very similar in characteristics to those seen in normoxia (e.g., rate of rise and fall as well as duration) can be generated for at least 3 h of anoxia in isolated (cortical slice) turtle cortical pyramidal cells.[5]

In the goldfish, Suzue et al.[27] and Fay and Ream[6] have shown that anoxia and asphyxia cause a strong suppression of sound-evoked electric activity in the auditory system, and they attributed this depression to a lowered presynaptic activity and a decreased excitatory transmitter release. Electrical activity is therefore suppressed during anoxia in the anoxia tolerant brain, but the capacity to respond is preserved.

4.4 Metabolic Rate

In the turtle, whole body heat production falls to 10-15% of aerobic values during 3 h of forced submergence, and in the goldfish the systemic energy consumption is reduced to one third of the normoxic rate after 3 h of anoxia.[32] The crucian carp is so well adapted to anoxia that its anoxic survival appears to be only limited by a complete depletion of its glycogen store – the largest found in any vertebrate, those that died during a 17 day anoxic period had run out of glycogen while survivors still had glycogen left (Fig. 6.4). Since glycogen depletion appears to be the only factor that normally limits the anoxia survival time in anoxia tolerant vertebrates, any reduction in the metabolic rate during anoxia, viz. the rate of glycogen use, will act to prolong anoxia endurance. However, the degree of metabolic depression that an anoxic tolerant animal can sustain will be a trade off between maximizing anoxia survival time and the necessity of sustaining the level of nervous and motor activity necessary to survive in the anoxic environment. In this requirement the crucian carp and turtle may differ. During anoxia the crucian carp, which remain active in ice covered anoxic ponds, shows a 50% reduction in spontaneous locomotor activity which by itself probably corresponds to a 35-40% reduction in energy use[19] while the anoxic turtle, which typically hibernates in aquatic burrows, is lethargic, showing little response to external stimuli. This difference may be reflected in the degree of metabolic depression reached by the anoxic turtle (producing heat at

1/10 of the normoxic rate) compared to the anoxic goldfish (1/3 of the normoxic rate), suggesting that the turtle has sacrificed behavioral responsiveness in favor of metabolic depression to a higher degree than the anoxia tolerant fish.

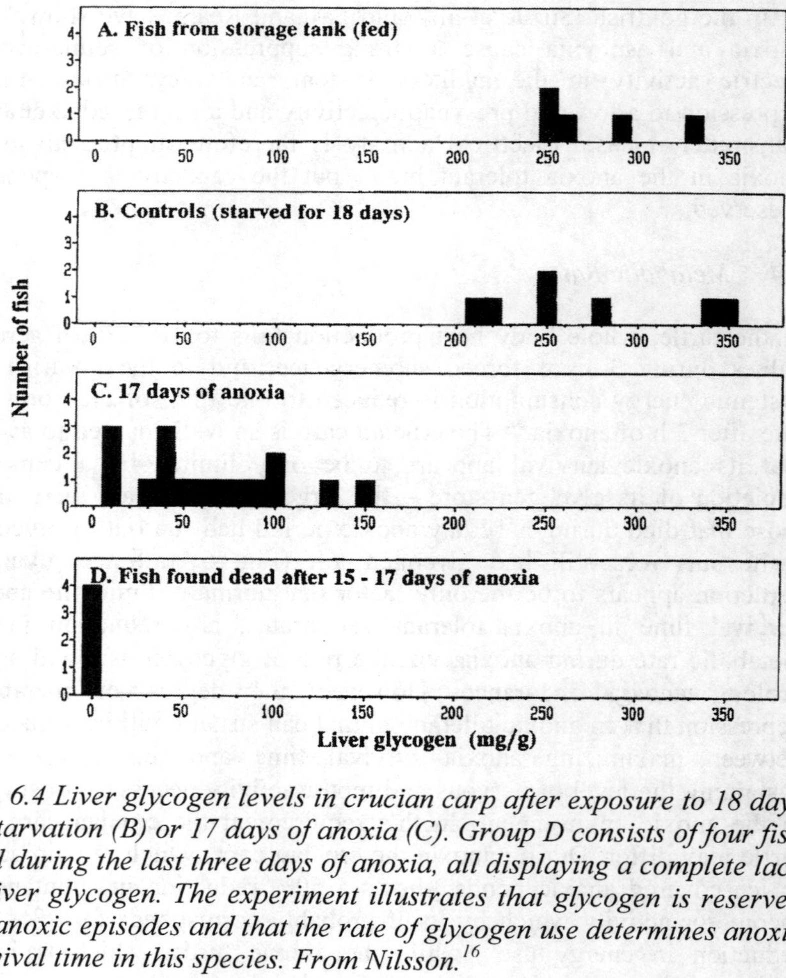

Fig. 6.4 Liver glycogen levels in crucian carp after exposure to 18 days of starvation (B) or 17 days of anoxia (C). Group D consists of four fish died during the last three days of anoxia, all displaying a complete lack of liver glycogen. The experiment illustrates that glycogen is reserved for anoxic episodes and that the rate of glycogen use determines anoxic survival time in this species. From Nilsson.[16]

As far as the brain is concerned calorimetric studies of isolated turtle brain slices indicate that anoxia produces a 40% decrease in the metabolic rate[20], and a similar decrease has been seen in the brain slice of the crucian carp[10]. These levels of metabolic depression in the isolated slices are likely to be a substantial underestimate of the true *in situ* value because the metabolic rate of brain slices is already about 50% reduced due to the absence of spontaneous activity.[12] An 80-90% reduction of brain metabolism in the turtle is a more likely figure that would agree with calculations on metabolite changes in the (*in vivo*) anoxic turtle brain.[3,14]

It would appear therefore, that in anoxia tolerant species anoxia results in a drastic depression of ATP turnover in the brain that at least matches a corresponding reduction in metabolic depression of the whole body. Only a few species appear to have the ability to substantially reduce the metabolic demand of the brain without compromising recovery. However brain hypometabolism is fundamental for their survival and the mechanisms by which this is achieved are central to the theme of this book.

5. REFERENCES

1. Blazka P. The anaerobic metabolism of fish. Physiol Zool 1958; 31:117-128.
2. Burggren WW, Schelton G. Gas exchange and transport during intermittant breathing in chelonian reptiles. J Exp Biol 1979; 82:75-92.
3. Chih CP, Feng ZC, Rosenthal M, Lutz PL Sick T. Energy metabolism, ion homeostasis, and evoked potentials in anoxic turtle brain. Am J Physiol 1989; 257:R854-R860.
4. Congleton JL. The respiratory response to asphyxia of *Typhlogobius californiensis* (Teleostei: Gobiidae) and some related gobies. Biol Bull 1974; 146:186-205.
5. Doll CJ, Hochachka PW, Reiner PB. Effects of anoxia and metabolic arrest on turtle and rat cortical neurones. Am J Physiol 1991; 260:R747-R755.
6. Fay RR, Ream TJ. The effects of temperature change and transient hypoxia on auditory nerve fibre response in the goldfish (*Carassius auratus*). Hearing Res 1992; 58:9-18.
7. Fernandes JA, Lutz PL, Tannenbaum A, Todorov, A.T. Liebovitch, L. Vertes R.. Electroencephalogram activity in the anoxic turtle brain. Am.J.Physiol; 1997;273, R 911-R919.
8. Jackson DC. Metabolic depression and oxygen depletion in the diving turtle. J Appl Physiol 1968; 24:503-509.
9. Jobsis FF. What is a molecular oxygen sensor? What is a tranduction process. In: Reivich M, Coburn R, Lahiri S, ed(s). Tissue hypoxia and ischemia. New York: Plenum,1977: 3-18.
10. Johansson D, Nilsson GE, Törnblom E. Effects of anoxia on energy metabolism in crucian carp brain slices studied with microcalorimetry. J Exp Biol 1995; 198:853-859.
11. Kelly DA, Storey KB. Organ specific control of glycolysis in anoxic turtles. Am J Physiol 1988; 255:R774-R779.
12. Lutz PL. Anoxic defense mechanisms in the vertebrate brain. Ann Rev Physiol 1992; 54:601-618.
13. Lipton P, Whittingham TS. Energy metabolism and brain slice function. In: Dingledine D ed(s). Brain Slices. New York: Plenum,1984: 113-153.
14. Lutz PL, McMahon P, Rosenthal M, et al. Relationships between aerobic and anaerobic energy production in turtle brain in situ. Am J Physiol 1984; 247:R740-R744.

15. Lutz PL, Rosenthal M, Sick TJ. Living without oxygen: turtle brain as a model of anaerobic metabolism. Mol Physiol 1985; 8:411-425.
16. Nilsson GE. Long term anoxia in crucian carp: changes in the levels of amino acid and monoamine neurotransmitters in the brain, catecholamines in chromaffin tissue, and liver glycogen. J Exp Biol 1990; 150:295-320.
17. Nilsson GE, Lutz PL, Jackson TL. Neurotransmitters and anoxic survival in the brain: a comparison between anoxia tolerate and anoxia intolerant vertebrates. Physiol Zool 1991; 64:638-652.
18. Nilsson GE, Pérez-Pinzón M, Dimberg K. Brain sensitivity to anoxia in fish as reflected by changes in extracellular potassium-ion activity. Am J Physiol 1993; 264:R250-R253.
19. Nilsson GE, Rosén P, Johansson D. Anoxic depression of spontaneous locomotor activity in crucian carp quantified by a computerized imaging technique. J Exp Biol 1993; 180:153-162.
20. Pérez-Pinzón M, Bedford J, Rosenthal M, et al. Metabolic adaptations to anoxia in the isolated turtle cerebellum. Soc Neurosci 1991; 17:1269 (abst.).
21. Pérez-Pinzón M, Rosenthal M, Lutz PL, et al.. Anoxic survival in the isolated turtle cerebellum. J Comp Physiol 1992; 16:345-351.
22. Piironen J, Holopainen IJ. A note on seasonality in anoxia tolerance of crucian carp *(Carassius carassius (L.))* in the laboratory. Annal Zool Fen 1986; 23:335-338.
23. Robin ED, Lewiston N, Newman A.. Bioenergetic pattern of the turtle brain and resistance to profound loss of mitochondrial ATP generation. Proc Natl Acad Sci USA 1979; 76:3922-3926.
24. Rosenthal M, Sick TJ. Brain oxygenation: monitoring techniques and insights into brain function and survival. Bailliere's Clinical Anaesthesiology 1989; 3:647-674.
25. Sick TJ, Rosenthal M, Lemanna JC, Lutz PL. Brain potassium ion homeostasis, anoxia, and metabolic inhibition in turtles and rats. Am J Physiol 1982; 243:R281-R288.
26. Sick TJ, Lutz PL, LaManna JC, Rosenthal M.. Comparative brain oxygenation and mitochondrial redox activity in turtles and rats. J Appl Physiol 1982b; 53:1354-1359.
27. Suzue T, Wu G-B, Furukawa T. High susceptibility to hypoxia of afferent synaptic transmittion in the goldfish sacculus. J Neurophysiol 1987; 58:1066-1079.
28. Ultsch GR. Ecology and physiology of hibernation and overwintering among freshwater fishes, turtles and snakes. Biol Rev 1989; 64:435-516.
29. van der Boon J, De Jong RL, van den Thillart GE, Addink A . Reversed-phase ion-pair HPLC of purine nucleotides from skeletal muscle, heart and brain of the goldfish, *Carassius auratus* L. – II. Infuence of environmental anoxia on metabolite levels. Comp Biochem Physiol B 1992; 101:583-586.
30. Van Ginneken V, Nieveen M, Van Eersel R, Van den Thillart G, Addink A: Neurotransmitter levels and energy status in brain of fish species with and without the survival strategy of metabolic depression. Comp Biochem Physiol A 1996; 114:189-196.
31. Van Waarde A, Van den Thillart G, Kesbeke F. Anaerobic energy metabolism of the european eel, *Anguilla anguilla* L. J Comp Physiol 1983; 149:469-475.
32. Van Waversveld J, Addink ADF, Van den Thillart G. Simultaneous direct and indirect calorimetry on normoxic and anoxic goldfish. J Exp Biol 1989; 142:325-335.
33. Wegener G, Krause U. Environmental and exercise anaerobiosis in frogs. In: Hochachka PW, Lutz PL, Sick T, Rosenthal M, Van den Thillart G. ed(s). Surviving Hypoxia: Mechanisms of control and adaptation. Boca Raton: CRC Press,1993: 217-236.
34. Yamada S, Brauer F, Knierim D, et al. Can somatosensory evoked potential monitoring predict energy dynamics during controlled hypotension? Neurol Res 1992; 14:325-329.

CHAPTER 7

MECHANISMS OF BRAIN ANOXIA TOLERANCE

1. INTRODUCTION

So what strategies and adaptive mechanisms are used by anoxia-tolerant vertebrates to survive brain anoxia? As we have just emphasized, one primary strategy is to maintain brain ATP levels by drastically decreasing the rate of ATP consumption. One must distinguish here between energy production and demand. A depressed energy production per se, which is displayed by virtually all animals exposed to anoxia, including those that die within a few minutes, does not promote anoxic survival unless it is preceded by a corresponding decrease in energy demand. It is clear that in order to remain viable and maintain ionic and energy balance during anoxia, the very substantial reduction in brain cell energy expenditure must be matched by a corresponding reduction in energy demand. The mechanisms that produce these energy savings are the keys to brain anoxic survival. However, the anoxic brain must also be protected from the other sources of related brain injury, including excessive acidosis, the uncontrolled release of excitatory amino acids and the formation of free radicals (see Chapter 4).

As we will see, it appears likely that metabolic depression is essential for long-term anoxic survival since it saves the limited internal glycogen stores, glycogen being the only usable fuel during anoxia.

On the other hand, during the transition to the anoxic state it is probably essential to increase glycolytic ATP production to meet the rate of ATP use before the mechanisms of metabolic depression have come into play.

We will begin this chapter by discussing the adaptations for this "transition" glycolytic activation. The major part of the chapter will focus on mechanisms underlying the metabolic depression, and, finally,

---

The Brain Without Oxygen. Third Edition. Peter L. Lutz, Göran E. Nilsson and Howard M. Prentice.

we will point out that there are not only similarities, but also some striking differences in the way freshwater turtles, on one hand, and *Carassius*, on the other hand, have solved the problem of anoxic survival.

2. ENERGY PRODUCTION

2.1 Dependence on Glycolysis

The importance of glycolytic ATP production for anoxic survival of crucian carp as well as turtles is illustrated by experiments showing that these animals rapidly lose brain ion-homeostasis during anoxia (as seen by a drastic increase in the extracellular K^+ levels and depolarization) if the brain is superfused with the glycolytic inhibitor iodoacetate.[21,31,64,158] During normoxia, iodoacetate has no apparent effect on the extracellular K^+ concentration in brain, and as mentioned above, freshwater turtles and crucian carp normally maintains a low extracellular K^+ level in brain for many hours (and probably many days) of anoxia. The vulnerability of turtle brain to ischemia, further indicates that its tolerance of anoxia depends upon the continued provision of substrates through the circulation.[159]

Other evidence on the necessity of glycolytic activation for anoxia tolerance comes from experiments with the isolated freshwater turtle cerebellum. During anoxia, the glucose level in the surrounding Ringer solution must be increased from 5 mM to 20 mM to assure survival of this preparation, otherwise, a loss of evoked potentials, a fall in ATP levels and a general depolarization will soon occur.[136]

2.2 Glycogen Stores

A striking quantitative difference between anoxia-tolerant and anoxia-intolerant vertebrates is seen in the size of the glycogen stores. The genus *Carassius* and the freshwater turtles appear to have the largest liver glycogen stores of any vertebrates[54] in that glycogen stores can make up about 30% of the crucian carp liver[60] and 15% of the freshwater turtle liver.[23]

In brain, the glycogen reserves are almost exclusively localized in glia cells (astrocytes)[173], which corresponds to their role of nutritional support for neurons. In crucian carp, goldfish and freshwater turtle, the concentration of glycogen in brain range between 12.8 and 19.5 μmol hexose/g, while the brain glycogen contents of anoxia sensitive species like the rat, mouse and rainbow trout are much lower: only 2.2-3.7 μmol/g (Fig. 7.1). Schmidt and Wegener[156] found that in the crucian carp, the brain glycogen level falls by 75 and 95% during the first 2 and 7 h of anoxia, respectively. At the same time, there is an increase in the content of the active form (a-form) of the glycogen degrading enzyme glycogen phosphorylase in brain of goldfish and crucian carp.[14,156,168] The importance of the brain glycogen stores during the initial phase of anoxia in *Carassius* is further emphasized by the astonishing fact that, at room temperature, the level of ATP in goldfish brain is kept constant for 30 min after decapitation.[104]

The brain glycogen is probably responsible for brain glucose supply during the initial phase of anoxia, while the huge liver glycogen store takes over this responsibility after brain glycogen reserves have been depleted. Indeed, the exhaustion of the liver glycogen store appears to be the factor that finally puts a limit to anoxic survival.[114] Interestingly, the glycogen stores of the liver and other organs of the freshwater turtle are unchanged over 22 weeks of fasting at 20-22 °C[135], and there is no change in the liver glycogen store of crucian carp during 18 days of starvation at 8 °C (Fig. 6.4 in Chapter 6), indicating that glycogen is dedicated to provide energy substrate only during anoxia in these species.

2.3 Delivery of Glucose from Systemic Sources

In order for the brain to profit from the glucose stored in the liver, the delivery rate of glucose to the brain has to increase. This can be done in two ways. Firstly, the concentration of glucose in the blood can be elevated, and secondly, the rate of blood supply to the brain can be increased. Both these mechanisms seem to be utilized by anoxia-tolerant vertebrates.

2.3.1 Increased Blood Glucose Level

In goldfish the blood glucose concentration can increase 5-fold during anoxia[157,183], and in crucian carp anoxia produces a 2.7 fold increase in the whole body content of glucose[70]. In freshwater turtles (*Chrysemys* and *Trachemys*) held at room temperature, anoxia also produces a marked hyperglycemia, with plasma glucose rising from 3 mM to as much as 25 mM.[78,134]

Interestingly, such increases in blood glucose do not occur in anoxic freshwater turtles that are kept at low temperatures (5 °C).[79] One possible explanation is that during anoxia, glucose release by the liver is more tightly matched with tissue glucose consumption at low temperatures than at high temperatures. Freshwater turtles are probably better adapted to survive anoxia at low temperatures, since, in nature, these animals are most likely to encounter anoxia during their winter hibernation in ice-covered ponds and lakes. Consequently, anoxia is likely to be a much more stressful condition at high than at low temperatures.

Catecholamines may play a crucial role here. In turtles acclimated to room temperature, anoxia provokes marked increases in plasma catecholamine concentrations[78,184], the increases appearing to correlate with a fall in plasma pH.[185] It has been suggested that anoxia and acidosis act synergistically to stimulate the release of catecholamines[79], the increased catecholamines functioning to stimulate liver glycogen phosphorylase activity and thereby increase plasma glucose levels.[77]

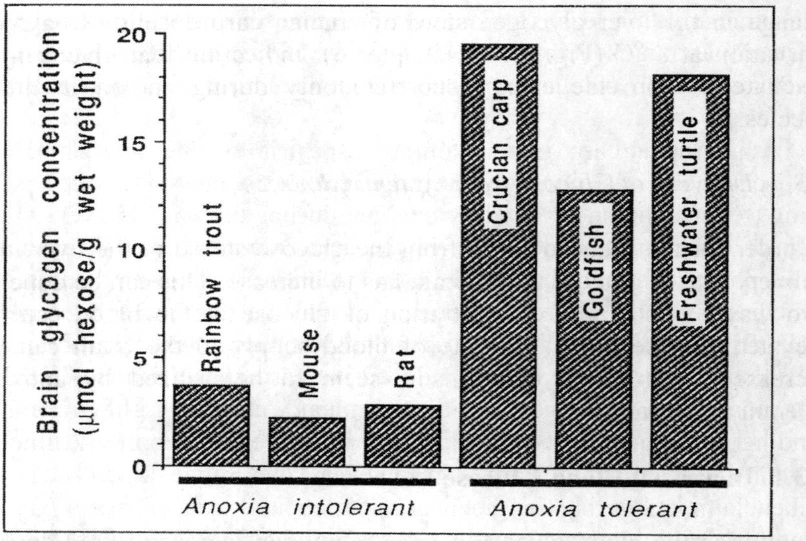

Fig. 7.1 Brain glycogen concentrations in anoxia-intolerant and anoxia-tolerant vertebrates. Values are from Schmidt and Wegener[156], McDougal et al.[104], Dunn and Hochachka[34], Lowry et al.[91] and Hawkins and Mans.[48]

By contrast, in the mammal elevated glucose levels are associated with increased brain damage in the hypoxic-ischemic brain. This is thought to be the result of increased lactate production and brain acidosis[142,164], although the issue is somewhat complex (see Chapter 4).

2.3.2 Increased Brain Blood Flow

In anoxia-tolerant vertebrates, the elevated blood glucose levels seen during anoxia are accompanied by an increase in brain blood flow. In crucian carp, a study utilizing epi-illumination microscopy, which allows a continuous monitoring of the brain blood flow, has shown that there is a doubling of the blood flow velocity during anoxia which was sustained for the whole duration of a 6 h anoxic exposure (Fig. 7.2).[121] The same study showed that a similar increase in blood flow is produced by topical application of adenosine during normoxia. Moreover, superfusing the crucian carp brain with the adenosine receptor blocker aminophylline was found to inhibit the effect of anoxia on brain blood flow, clearly suggesting a mediatory role of adenosine in the anoxia induced increase in brain blood flow displayed by the crucian carp.

In freshwater turtles, while initial studies agreed that there was an increase in brain blood flow following anoxia, they disagreed sharply in the magnitude of the increase as measured by different techniques. Davies[27,28,29], using labeled microspheres, found a 3-fold increase in brain blood flow after 15-30 min of anoxia, while, Bickler[7], who estimated the average blood flow from H_2 clearance between 0.5 and 3 h of anoxia, found a relatively modest (45%) increase. The discrepancies were later explained when brain blood flow was monitored continuously. A study using epi-illumination microscopy showed that the change in brain blood flow in the anoxic turtle brain is a dynamic event.[57] Anoxia causes an initial temporary increase in brain blood flow rate, reaching about 200% of the control value, whereupon flow rates fall back to normoxic values after about 115 min in anoxia (Fig. 7.3). The same study showed that similar to the crucian carp, adenosine is a potent stimulator of brain blood flow in turtles, and the anoxic increase in brain blood flow can be completely blocked by the adenosine antagonist aminophylline. Thus, adenosine is probably a main mediator of increased cerebral blood flow in anoxic turtles. It would appear then that the initial temporary increase in brain blood flow in the freshwater turtle brain fits well with the contemporaneous rise and fall in extracellular adenosine (see below).

Subsequent studies on crucian carp[56] and turtles[58] have indicated that NO is an endogenous vasodilator in the brain of these species, since NOS inhibitors (L-NA or L-NAME) are able to block the increase in brain blood flow induced by acetylcholine. The same studies showed that NOS inhibitors are without effect on the anoxic increase in brain blood flow in crucian carp and turtles. This is less surprising in the light of the fact that the NOS catalyzed production of NO from L-arginine demands molecular oxygen.[89] Thus, endogenous NO can hardly be a vasodilator under completely anoxic conditions. However, an interesting finding in the turtle was that also the NO donor SNP lost its ability to stimulate brain blood flow when the animal was exposed to anoxia[58], suggesting that the brain vasculature of the turtle somehow becomes unresponsive to NO during anoxia. Keeping in mind the numerous detrimental effects that NO appears to have during periods of neural energy deficiency (discussed in Chapter 4), it is tempting to hypothesize that the neural responsiveness to NO is somehow down-regulated in an animal that is so well adapted to anoxia as the freshwater turtle.

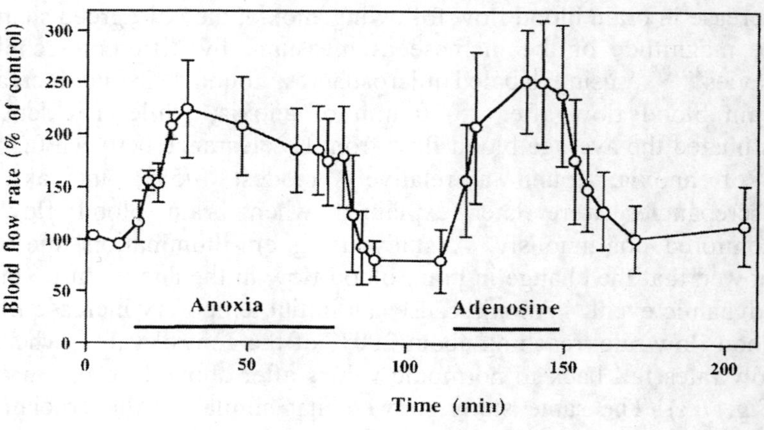

Fig. 7.2 Effects of anoxia and adenosine on brain blood flow rate in crucian carp. Blood flow rate was measured with epi-reflection microscopy on venules at the surface of the optic lobes. Values are means ±SD from 3 vessels. Modified from Nilsson et al.[121]

In turtles, the blood flow to most organs other than brain is substantially decreased during anoxia[27] (Fig. 7.4). This is clearly a mechanism aimed at reducing the cardiac work load and simultaneously allowing an increased brain blood flow. This redistribution of systemic blood flow, involving a reduced flow to most organs except the CNS, is strikingly similar to the mammalian diving response, displayed, for example, by diving seals[195] (see Chapter 8). A fundamental difference is of course that in the diving mammal, oxygen is still delivered by the blood, so brain blood flow is maintained rather than increased. In the anoxic crucian carp and turtle, the role of brain blood flow must primarily be to bring more glucose in and lactate out of the brain, and it has therefore to be increased, at least initially (before metabolic depression sets in). The fact that the turtle down regulates (or merely maintains) the blood flow to non-neural organs[27] indicates that oxygen deficiency is primarily a problem for the CNS in this species. In other words, when oxygen disappears, other organs do not need the special energetic priority that the brain requires. In turtles, glycolytic activation is attenuated after an hour or so of anoxia, and glycolytic downregulation is increasingly expressed as the rate of energy consumption becomes more and more depressed.[80] In the anoxic turtle brain, key enzymes associated with glycolysis are in less active forms at the time when glycolytic rates are depressed, as indicated by changes in the levels of glycolytic intermediates.[80] Phosphofructokinase in particular shows increased inhibition by ATP and citrate, and a decreased sensitivity to its activator AMP, during this period, and glycogen phosphorylase-a activity is reduced by 70%[13], indicating a decrease in the breakdown of glycogen to glucose.

2.4 Glycolytic Adaptations

In the turtle, the covalent modification of key glycolytic enzymes by phosphorylation may play a role in coordinating the depression of glycolysis by producing less active forms of the enzymes.[34,171] The fall in pH that accompanies prolonged anoxia can also act to depress enzyme activities, particularly those associated with glycolysis.[52]

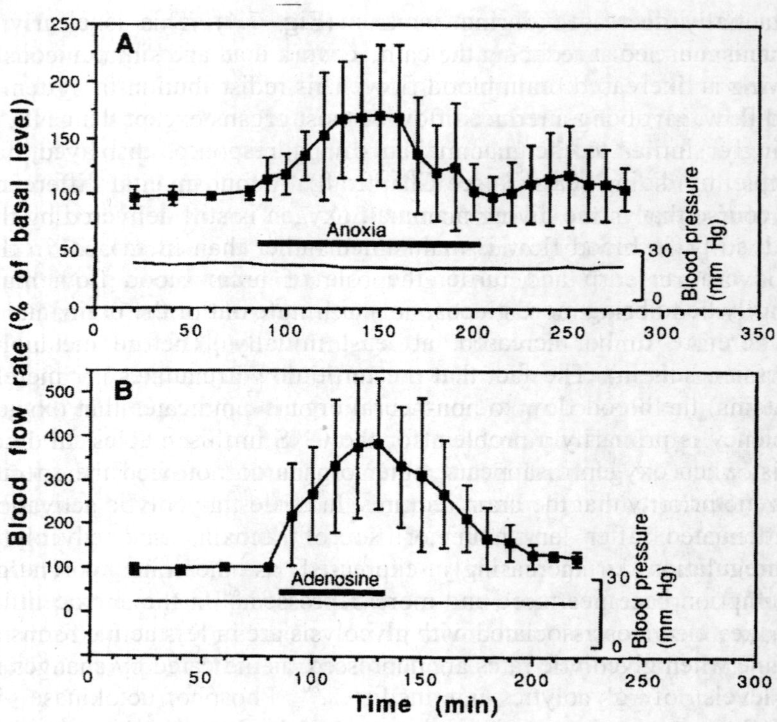

*Fig. 7.3 Effects of anoxia and adenosine on brain blood flow rate in the freshwater turtle (*Trachemys scripta*). Blood flow rate was measured with epi-reflection microscopy on venules at the surface of the optic lobes. Values are means ±SD from 4-7 animals. Modified from Hylland et al.*[57]

An interesting parallel to the biphasic increase and decrease in the glycolytic rate of the turtle brain has been observed in guinea pigs. Here, the glycolytic rate in synaptosomes is initially stimulated 8-10 fold by anoxia. However, this stimulation is transient, and after 15 min, glycolytic rates decline toward the pre-anoxic rates.[76] The fall in glycolysis is Ca^{2+} dependent, and it is not seen when Ca^{2+} is absent, suggesting to the authors that elevated Ca^{2+} is the cause of glycolysis inhibition rather than energy failure per se.[76] The rapid decline in anoxic glycolysis, which is linked to an increase in cytosolic Ca^{2+}, is a time limiting factor for anoxic survival in the mammal.

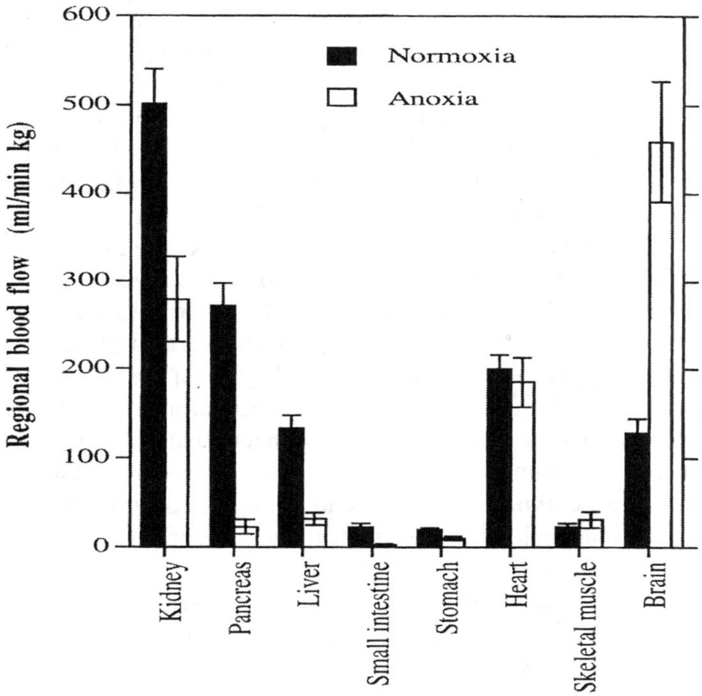

Fig. 7.4 Changes in the distribution of systemic blood flow in the freshwater turtle (Trachemys scripta) induced by 30-45 min of nitrogen respiration. Blood flow was measured with the radioactive microsphere technique. From Davies.[27]

In the anoxic turtle, although glycolysis is severely depressed, it is still the only source for ATP production and lactate therefore accumulates during anoxia. The amounts can be very substantial, e.g. plasma lactate concentrations as high as 200 mM have been seen in the freshwater turtle *Chrysemys picta*, after 3 months submergence in cold (3 °C) anoxic water, accompanied by a almost 25-fold increase in the Ca^{2+} level in blood plasma.[175]

2.4.1 Ethanol: The Alternative Glycolytic End-Product

The *Carassius* species, the crucian carp as well as the goldfish, have the unique ability to produce ethanol as the major metabolic end product during anoxia.[70,147] By producing ethanol, the crucian carp avoids the build up of high lactate levels and consequent acidosis. Clearly this is a systemic adaptation that will also facilitate anoxic survival of the brain.

The ethanol producing pathway of *Carassius* is summarized in Fig. 7.5. The unique part of the pathway starts with pyruvate, which is either produced from the normal equilibrium reaction from lactate (catalyzed by lactate dehydrogenase), or it is directly derived from the glycolysis. Inside mitochondria, pyruvate is then decarboxylated to form acetaldehyde and CO_2 in a reaction catalyzed by pyruvate dehydrogenase. This enzyme occurs in all vertebrates as a part of the pyruvate dehydrogenase complex, which converts pyruvate to acetyl-coenzyme A for delivery to the citric acid cycle. In all other vertebrates, however, acetaldehyde is only an intermediate in the reactions catalyzed by this enzyme complex, and acetaldehyde is never allowed to leave the complex. However, in *Carassius*, the pyruvate dehydrogenase complex seems to start "leaking" during anoxia, at least in muscle tissue, allowing the intermediate acetaldehyde to slip out.[80] Acetaldehyde then diffuses out of the mitochondria and once out of the cytosol, it is rapidly converted to ethanol by alcohol dehydrogenase (ADH) – an enzyme present at extremely high levels in the skeletal muscles of *Carassius.*[112,116,157] Indeed, ethanol appears to be produced solely in skeletal muscle (red as well as white), since this is the only tissue where high activities of alcohol dehydrogenase are found. Other tissues in *Carassius*, including the brain, lack ADH[112,157], so lactate must still be the major anaerobic end product in brain. Indeed, when brain slices from crucian carp are exposed to anoxia, only lactate, and no ethanol, is released to the bathing solution.[68] The most likely reason why the crucian carp brain does not produce ethanol is that, like all vertebrate brains, the crucian carp brain contains relatively high activities of aldehyde dehydrogenase (ALDH)[112,116], which catalyzes an important step in the degradation of monoamine neurotransmitters (see chapter 2). ALDH has a high affinity for acetaldehyde (higher than ADH), so if ALDH were present in an ethanol producing tissue, there would be a competition between ALDH and ADH for acetaldehyde. Hence, much of the acetaldehyde would be transformed into acetic acid by ALDH rather than to ethanol. The overall result is that the lactate produced by the brain must be transported by the blood to the muscles to be transformed to ethanol and CO_2. Ethanol readily penetrates membranes, and ethanol released to the blood will leave the fish through the gills.

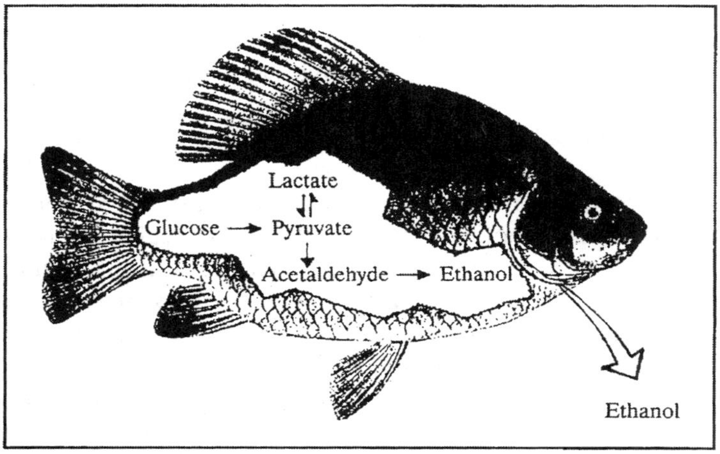

Fig. 7.5 Ethanol producing pathway in anoxic crucian carp. Together with the congeneric goldfish, the crucian carp has the unique ability to produce ethanol as the major anaerobic endproduct. By releasing ethanol into the water, a buildup of the anaerobic endproduct is avoided.

It might be expected, of course, that the blood ethanol in anoxic *Carassius* would affect brain function. The blood ethanol levels (ca 5 mM) are unlikely to be high enough to directly reduce nervous activity and to have any direct anesthetic action. But the ethanol appears to have a particular role in facilitating metabolic depression and may act as a messenger of the anoxic conditions. Thus, when keeping goldfish in a temperature gradient, Crawshaw et al.[25] found that intracranial injections of a minute amount of ethanol (47.5 pg) into the nucleus preopticus periventricularis (NPP) made the goldfish select a water temperature of 16 °C, instead of the 26 °C that was normally preferred. These results suggested that the ethanol produced during anoxia functions as a signal for moving to cooler water when it enters the NPP. Indeed in goldfish as well as other vertebrates NPP is thought to control behavioral temperature regulation. However, a more recent study by the same group[144] suggests that ethanol is probably not involved in behavioural temperature regulation in *Carassius*: When faced with falling ambient oxygen levels, goldfish start moving to cooler water well before any endogenous ethanol is produced.

It should be mentioned that behavioral hypothermia in hypoxia has been found in many animals, from paramecium to reptiles.[188] The obvious advantage of being in cooler water during anoxia being a lower metabolic rate, and, hence, prolonged anoxic survival time.

3. METABOLIC DEPRESSION

Metabolic depression is the downregulation of energetic flux. The most important aspect of metabolic depression is the downregulation of ATP consumption since a decrease in ATP production by itself would, of course, only lead to a fall in ATP levels with consequent death. In turtles, the down regulation in the activity of certain glycolytic enzymes must be a secondary adjustment aimed at matching the rate of ATP production with a new state of depressed ATP consumption. Thus, to be compatible with anoxic survival, glycolytic depression will follow the primary adaptation, the decrease in the use of ATP. This section will therefore focus on mechanisms to reduce the rate of ATP utilization.

In anoxia-tolerant animals, metabolic depression is produced through behavioral as well as cellular adjustments in the rate of energy use. One energy saving behavioral adjustment displayed by *Carassius* has already been mentioned: the tendency to move to cooler water when encountering anoxia. In addition to this, both *Carassius* and freshwater turtles display a lowered physical activity during anoxia. The anoxic crucian carp reduces its spontaneous locomotor activity to about half that seen during normoxia (Fig. 7.6).[127] A more striking reduction is seen in freshwater turtles, which at room temperature, appear completely limp within 30-60 min of anoxia, (which is approximately the time it takes them to deplete their internal oxygen stores[94]). That the brain controls the induction of this inactive state is clearly demonstrated by the fact that, in contrast to whole animals, bodies of decapitated turtles (*Trachemys*) continue to move for more than 24 h, although they are obviously completely anoxic (Lutz and Nilsson, unpublished observations).

Measurements of the heat production of anoxic goldfish and turtles also reveal differences. Anoxic goldfish depress their whole body rate of heat production to about 1/3 of that seen during normoxia[180], whereas anoxic turtles lower their heat production to 1/10 of the normoxic value.[57]

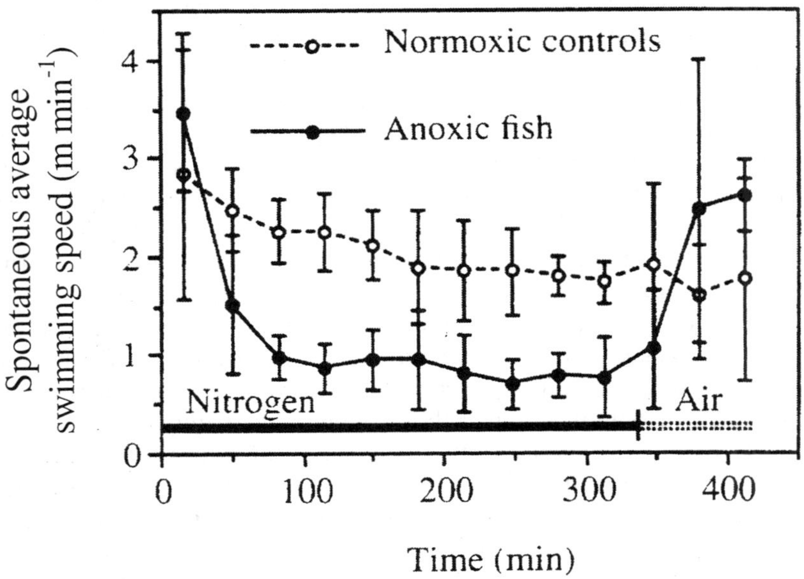

Fig. 7.6 Anoxic depression of spontaneous locomotor activity in crucian carp (kept at 9 °C). The locomotor activity (swimming speed) was measured with a computerized imaging technique. From Nilsson et al.[127]

With regard to the brain, calculations based on changes in energy metabolites indicate that the anoxic turtle brain reduces its metabolic rate to around 1/20 of the normoxic rate.[19,97] Since anoxic *Carassius* remain responsive, it is highly unlikely that their brains are so deeply depressed. Microcalorimetric measurements of heat production of crucian carp brain slices show a 31% reduction in ATP use in response to anoxia.[68] However, measurements on isolated brain slices are likely to substantially underestimate the *in situ* metabolic depression because brain slices are already metabolically depressed.[90] Anoxic turtle brain slices only show a 40% reduction in metabolic rate.[33]

4. ELECTRICAL ACTIVITY

Metabolic depression in brain is revealed by a suppressed electrical activity. The resultant energy savings from the brain becoming quiescent may be considerable. As discussed in chapter 2, electrical activity is probably responsible for at least 50% of the energy consumption of the mammalian brain and in the mammalian brain, an early depression of electrical activity following ischemia/anoxia is thought to provide some (albeit slight) anoxia tolerance by, for example, delaying the increase in extracellular K^+ activity that ultimately results in anoxic depolarization.[143] Electrical activity may account for an even greater proportion of the energy consumption of the ectothermic brain since this organ spends less energy in counteracting the ion-leak fluxes that underlie heat production in endotherms.

The EEG of normal air breathing turtles has a relatively low amplitude background but has a wide spectrum of frequencies comparable to those seen in mammals.[42] During the transition period to anoxia there is a drastic fall in brain electrical activity[42,158] (Fig. 7.7) and the root mean square (RMS) voltage of the EEG reaches a depressed floor activity level of about 20% of normoxia after about 100 min. Interestingly, during the transition period, bursts of high voltage rhythmic slow (~3-8 Hz) waves appear, but subsequently decline and are absent during prolonged anoxia.[42] A rhythmic slow bursting activity in the EEG of mammals and birds has been connected to reduced activity during sleep and estivation.[105,182] In the turtle the pattern of EEG changes seen during the transition period indicates more than a mere global depression of electrical activity; it suggests a systematic and structured shut down of some of the electrical components of the brain.[42] Little further depression in electrical activity is seen after the full anoxic state is reached, but this basal anoxia EEG is periodically (every 0.5-2 min) interrupted by short (3-15s) bursts of mixed frequency activity.[42]

The hibernating mammal also shows distinct neuronal firing patterns during deep hibernation, primarily slow wave activity is periodically interrupted by short bursts of low amplitude activity that underlie arousal states of mammals.[83] This pattern of EEG activity, seen in both the severely metabolically depressed anoxic turtle and hibernating mammal, may represent the minimal or basic electrical activity that is compatible with the survival of the brain as an integrated unit, and allows the brain to return to normal functioning as aerobic metabolism is resumed. In addition to the changes in EEG pattern, evoked field potentials and excitatory postsynaptic potentials (EPSP) are also depressed in the anoxic turtle brain to about 50-80% of normoxic levels, indicating a corresponding reduction in synaptic transmission (Fig. 7.8).[39,40,136,137] With regard to the crucian carp CNS, light evoked potentials in the visual system has been studied. In these experiment's, the potentials evoked by light directed against the eye were measured at the level of the retina (electroretinogram, ERG) as well as in the optic tectum. The results showed that anoxia rapidly cause about an 80% suppression of the light evoked potentials in both the retina and the optic tectum (Fig. 7.9), suggesting that a down regulation occurs at the level of the retina.[69]

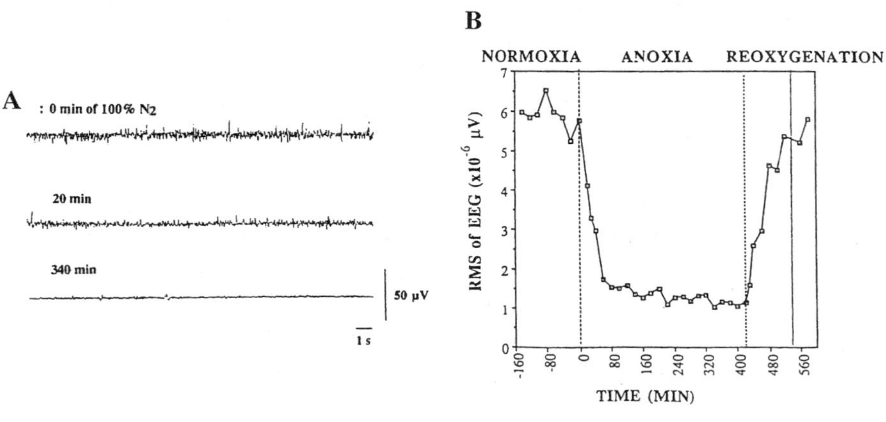

Fig. 7.7 A *The effect of anoxia on the EEG from the turtle cortex.* B *The root mean square voltage(RMS) of the turtle EEG as a function of time in anoxia and recovery. From Fernandes et al.*[42]

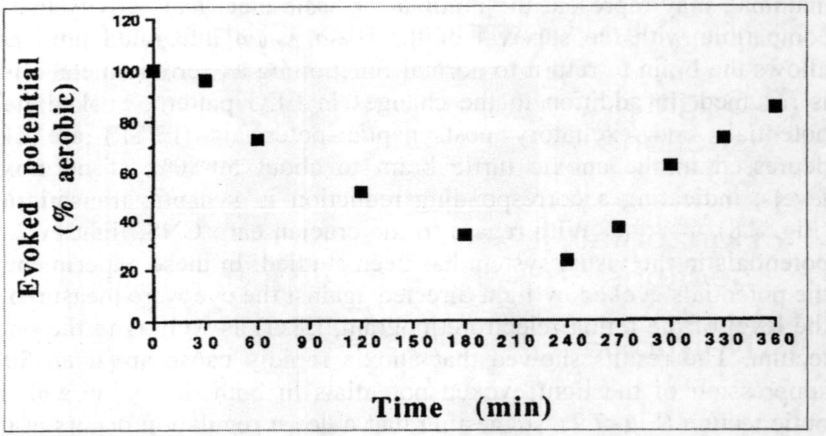

Fig. 7.8 Effects of anoxia (N_2 respiration at time 0) and recovery (air respiration at 240 min) on evoked potential amplitudes in turtle brain (olfactory bulb). Adapted from Feng et al.[40]

Similarly, sound-evoked electric activity in the sacculus and auditory nerve have been found to be depressed in anoxic goldfish.[38,172] The sacculus and auditory nerve are (in contrast to the retina) parts of the peripheral nervous system (PNS) rather than the CNS. Thus, in *Carassius*, there are evidence for electrical suppression both in the CNS and PNS.One of the major factors that influence post synaptic responsiveness is the electrical resistance of cell membranes. Adequate membrane resistance assures that postsynaptic currents are not dissipated and that the amplitude of postsynaptic potentials are maintained for substantial distances from the site of synaptic contact. Action potentials would be inhibited if resting membrane potentials were either hyperpolarized further from the action potential firing threshold or were clamped due to the opening of K^+ or Cl^- channels. Thus a decrease in input resistance involving K^+ or Cl^- channels could depress synaptic transmission by short circuiting synaptic potentials. As discussed before, a rapid decrease in input resistance is an early response of many mammalian neurons during anoxia. The major mechanism appears to be an activation of specific K^+ conductances[84], with the resultant hyperpolarization inhibiting neuronal firing.

Like mammals, anoxia produced a decrease in membrane input resistance in the isolated cerebellum of the turtle *Trachemys scripta.*[137] But unlike the mammal, was no evidence of hyperpolarization in the anoxic turtle cerebellum[137] and the membrane potential remained constant during anoxia in cortical sheets from the closely related turtle *Chrysemys picta.*[16] No changes in membrane resistance were seen during anoxia in pyramidal cells from telencephalic slices of this species.[31] However, a significant decrease in whole cell neuronal conductance has been recorded in the anoxic turtle cortex.[45] Membrane hyperpolarization, which provides minutes of protection, is clearly not essential for anoxic survival since both the isolated turtle cortex and cerebellum can endure hours of anoxia. Raising the action potential threshold could be a much more important strategy. This would depress the electrical activity of the neuron by increasing the synaptic input needed to trigger the action potential and therefore, decrease the rate of action potential firing. In the isolated turtle cerebellum Na^+-channel action potential thresholds are increased by as much as 14 mV during anoxia[137]. Buck and Bickler[16] found anoxia induced increases in the threshold potential in cortical sheets from the turtle *C. picta.* These increases are much greater than that seen in the anoxic rat brain (3-5mV, in hippocampal CA1 neurons[44]), suggesting that the mechanisms responsible for decreasing action potential thresholds may be enhanced in the turtle brain.

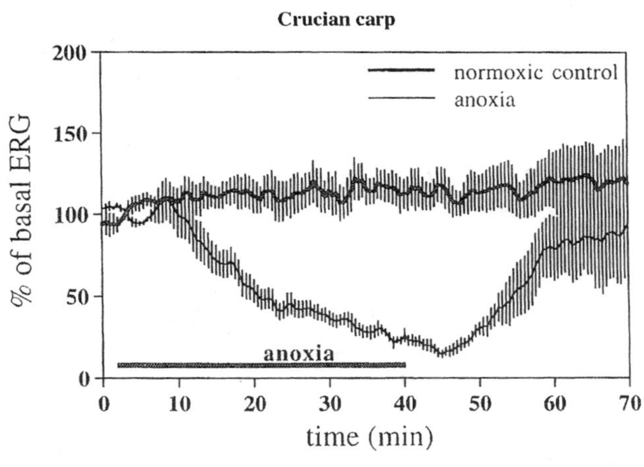

Fig. 7.9 Crucian carp becomes virtually blind in anoxia. The graph shows the reductions in the b-wave of the electroretinogram and evoked potential in the optic tectum in response to light exposure in a crucian carp during 85 min anoxia and 45 min normoxia. The eye was exposed to white light for 5 s every minute. From Johansson et al.[69]

The physiological and molecular processes causing the depression of electrical activity have long remained elusive. However, over the last years, a number of studies have shed some light on probable and possible mechanisms underlying metabolic depression. These mechanisms can be divided into two categories: (1) those that involve changes in the release of neurotransmitters/neuromodulators, and, thus, the activation of ligand-gated ion channels, and (2) those that depend on a change in the properties of other ion channels, such as the voltage-gated ion channels.

5. NEUROTRANSMITTERS AND NEUROMODULATORS

Changes in neurotransmitter release are of course of paramount importance in the regulation of brain activity. It is therefore likely that neurotransmitters and neuromodulators play a major role in the initiation and coordination of the processes that provide anoxic tolerance. An increased release of inhibitory neurotransmitters and neuromodulators could cause a depression of synaptic transmission and electrical activity. While excitatory neurotransmitters play a detrimental role in the anoxic mammalian brain, inhibitory neurotransmitters and neuromodulators may function to protect the anoxia-tolerant brain from energy failure, thereby promoting anoxic survival. Indeed, experimental evidence is now accumulating in support for such mechanisms.

5.1 Adenosine: a Signal for Energy Failure and Promoter of Metabolic Depression

All evidence points to adenosine playing a key role in brain anoxic survival. After retaliatory responses to hypoxia have failed (see chapter 4) cellular signals for an approaching energy failure are needed in order to initiate the metabolic adjustments required for anoxic survival. Although brain levels of ATP, ADP and AMP are maintained throughout many hours of anoxia, a limited but significant temporary fall in ATP (and ADP) is seen in the freshwater turtle brain early on, during the transition period of hypoxia to anoxia (Fig. 7.10).[80,97] Nilsson and Lutz[123] have suggested that this fall in ATP occurs in the turtle when the production of ATP from both aerobic (hypoxic) and anaerobic glycolysis is insufficient to meet the normal demand for ATP. They proposed that the ensuing fall in ATP and other phosphorylated adenosines results in a concomitant rise in adenosine, which then acts as a signal of energy insufficiency, and is directly involved in initiating the changes that result in stimulated glycolysis as well as in the drastic reduction of the metabolic rate. When the full hypometabolic state is established, anaerobic glycolysis, even at a depressed rate, is sufficient to meet the reduced ATP needs of the cell and its energy status is restored.

Endogenous adenosine is widely regarded as having an important neuroprotective role during ischemia and hypoxia in the mammalian brain.[151] In these situations adenosine is released into the extracellular space[181] of the brain where it acts to produce an increase in cerebral blood flow[24], stimulate glycogenolysis[102] and to decrease neuronal excitability as well as suppress excitatory neurotransmitter release.[141,167] It has been shown that adenosine delays the onset of harmful hypoxic depolarization in the rat hippocampus, acting via the A1 receptors.[87] Indeed adenosine has been termed a "retaliatory metabolite" for several tissues including the heart and brain since it is produced during conditions of energy insufficiency and subsequently acts to reduce energy consumption and increase energy supply.[111] Adenosine may also be involved in behavioral responses to hypoxia. In amphibians (toads) adenosine mediates a hypoxia induced movement to colder waters, which by lowering the metabolic rate would prolong anoxia survival.[12] In the turtle, there is a marked but temporary rise in extracellular adenosine after the onset of brain anoxia (Fig. 7.11) which is probably linked to the contemporaneous fall in ATP.[123] While the ability of adenosine to protect the anoxic mammalian brain is clearly limited (the survival time still being in the order of minutes versus weeks), in the anoxia tolerant animal the protective role of adenosine may be the first step in metabolic down regulation. Indeed, adenosine may even be involved in regulating the whole body metabolic depression that is characteristic of anoxia tolerant species. This is suggested by the observation that administration of the adenosine receptor blocker aminophylline caused the crucian carp to increase its anoxic metabolic rate, seen as an enhancement in the production of ethanol -the principal end product of anaerobic glycolysis for this species, while the normoxic metabolic rate (measured as O_2 consumption) was unaffected.[117] Adenosine also appears to be a key mediator of "channel arrest" playing a critical role in the anoxia induced decrease in turtle brain NMDA receptor activity[16] and potassium permeability.[132]

Adenosine may also promote anoxic survival of anoxia-tolerant vertebrates by stimulating increased cerebral blood flow. Evidence for such a mechanisms in crucian carp as well as in turtles was discussed earlier in this chapter.

The hypothesis that adenosine plays a critical role in anoxia tolerance is supported by the finding that superfusing the anoxic isolated turtle cerebellum with theophylline (an unspecific adenosine receptor blocker) or the specific adenosine A_1-receptor blocker 8-cyclopentyltheophylline (8-CTP) causes rapid depolarization.[139] In contrast, superfusing the anoxic *in situ* brain with the adenosine receptor antagonists (theophylline, 8-CTP) does not produce depolarization, although it does inhibit the full expression of channel arrest (see below).[132] The difference is probably due to the *in vitro* preparation being energy limited, depending on the diffusion of glucose through the slice to meet its energy needs, while in the *in situ* brain glycolytic delivery is greatly enhanced during anoxia through an increased brain blood supply and hyperglycemia.

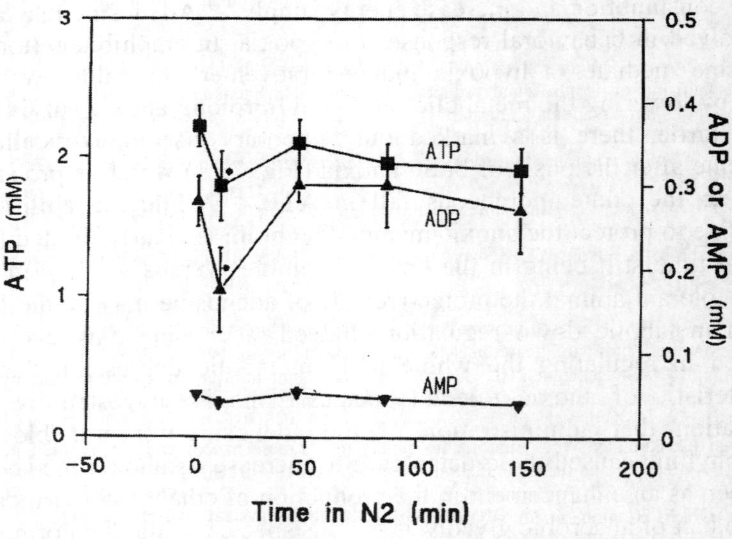

Fig. 7.10 Changes in the brain levels of ATP, ADP and AMP during N_2 respiration in the freshwater turtle. Adapted from Lutz et al.[97]

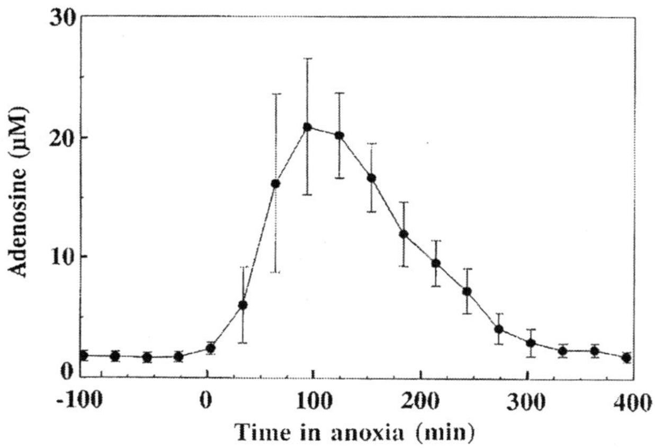

Fig. 7.11 Increase in extracellular adenosine during anoxia in the freshwater turtle brain (striatum) measured by intracerebral microdialysis. From Nilsson and Lutz.[123]

The role of adenosine as a metabolic depressor during periods of hypoxia may have a very ancient history in evolution. There is evidence of such an effect in hagfish, which belong to the most primitive group of vertebrates, the cyclostomes. Injecting the adenosine receptor blocker, theophylline, into hypoxic rainbow trout and hypoxic Pacific hagfish (*Eptatretus stouti*) resulted in significantly higher concentrations of plasma lactate and a more pronounced acidosis in both species, indicating an enhanced rate of glycolysis.[3] In the trout, there was also an increased rate of glycogen and creatine phosphate depletion in the heart. At the same time, theophylline was without effect on normoxic control fish. A parallel study using theophylline suggested a feed-back role for adenosine in depressing catecholamine release in response to anoxia in both the rainbow trout and the Pacific hagfish.[4] Adenosine also acts as a metabolic depressant in the marine worm *Sipunculus nudus*, an anoxia-tolerant invertebrate. In this species adenosine levels have been found to increase in the nervous system following exposure to either anoxia or hypercapnia – two conditions known to induce metabolic depression in *Sipunculus*.[145] The highest adenosine levels were found in worms exposed simultaneously to anoxia and hypercapnia. Moreover, injections of adenosine caused a marked depression of the oxygen consumption of control specimens. The experiments indicated that adenosine was acting through the nervous system of the worm rather than directly on the muscular body wall.

5.2 Inhibitory Neurotransmitters: Possible Consolidators of Metabolic Depression

5.2.1 Gamma-Amino Butyric Acid (GABA)

The tissue concentration of GABA, the most abundant inhibitory neurotransmitter in the vertebrate nervous system, increases considerably in anoxia-tolerant vertebrates in response to anoxia[126] (Fig. 7.12). For instance, in the turtle brain, GABA concentrations increase 45-60% after 2-4 h anoxia at 25 °C[50,95] and 127% after 13 h anoxia at the same temperature.[120] In the crucian carp, a 30% increase in the brain GABA levels is seen after 4.5 h at 18 °C[118], and after 17 days in anoxia at 8 °C, the brain GABA concentration has increased 5 times.[114]

An increase in the brain level of GABA has also been demonstrated in mammals and other anoxia-intolerant vertebrates during hypoxia and anoxia, and it is probable that this change occurs in most brains suffering low oxygen conditions.[125] In fact, GABA levels have been found to increase during anoxia in the brain of an invertebrate: the shore crab (*Carcinus maenas*).[124]

However, for anoxia-intolerant vertebrates such as mammals, the greatest increase in brain GABA will occur post mortem since they die so rapidly in anoxia. As discussed in chapter 2, the most likely reason why the GABA increases during hypoxia is that the conversion of glutamate to GABA is anaerobic and will continue during anoxia, while the breakdown of GABA requires oxygen because it needs two substances that are in short supply during anoxia, NAD and the tricarboxylic acid cycle intermediate, ά-ketoglutarate. However, neurotransmitters are only effective when they are released and can bind to their specific membrane receptors. Thus, changes in tissue concentrations of neurotransmitters becomes mechanistically relevant only if they are reflected in changes in the extracellular levels. Indeed, this appears to be the case. A study using in situ microdialysis in the striatum of the turtle brain has indeed shown that after about 100 min of anoxia at 25 °C (i.e. at about the time when extracellular adenosine is declining, see above), there is a large and sustained release of GABA, its extracellular level rising from 0.3 to 27 μM over a period of 150 min (Fig. 7.13). Thus, the rise in the extracellular GABA concentration exceeded the intracellular increase in GABA.

Microdialysis measurements in crucian carp[59] show that the brain (telencephalon) of this species also responds to anoxia by releasing GABA to the extracellular space. However, the degree of increase in extracellular GABA levels is much less, about 50% after 6 h of anoxia at 15 °C, compared to the almost 100 fold increase seen in the freshwater turtle brain. Clearly, this difference could underlie the great differences seen in the behavioural response of these two species to anoxia – being able to swim around like the anoxic crucian carp, or being totally comatose, like the anoxic turtle.

There is clear evidence that GABA has similar inhibitory actions in turtle and mammalian brain. Kriegstein and Connors[82] demonstrated the classical characteristics of interneuronal inhibition by GABA receptors in the *in vitro* dorsal cortex of the turtle. The early part of the inhibitory response was mediated by a Cl^- conductance presumably via the $GABA_A$ receptor, however, the later phase of the inhibition appeared to be mediated by a K^+ conductance, possibly coupled to $GABA_B$ receptors. On a much shorter time scale, Chen and Chessler[20] have shown that in the isolated turtle cerebellum, GABA produces an almost instantaneous extracellular alkaline transient through $GABA_A$ receptors gating an HCO_3^- efflux, and they suggested that this GABA evoked alkalization is a concomitant of inhibitory transmission.

As far as receptors are concerned, the B_{max} for the binding of [^{3}H]flunitrazepam, a specific agonist for the $GABA_A$ receptor, to the cerebral hemispheres of the control turtle (2404 fmol/mg protein) is within the range of those found for homologous regions of the rat brain (viz. rat cortex B_{max} = 2583 fmol/mg protein, rat striatum B_{max} =1886 fmol/mg protein[128]) indicating that the turtle brain has a similar density of $GABA_A$ receptors as the mammal.[99] The rat and turtle $GABA_A$ binding sites also show identical regulatory and kinetic properties in benzodiazepine binding.[153] The widely divergent turtle and mammal brains appear therefore, to have a similar GABAergic capacity.

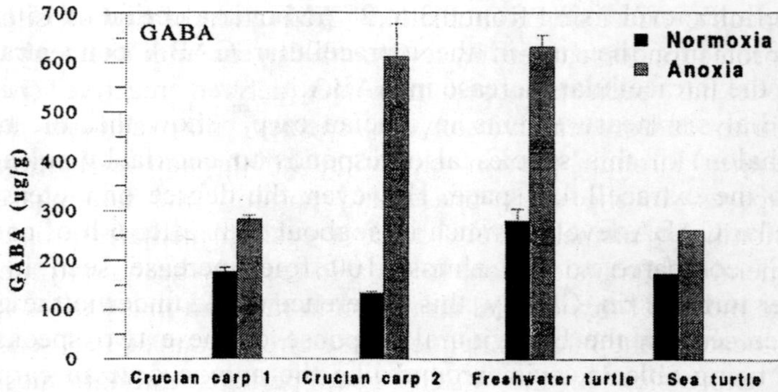

Fig. 7.12 Effects of anoxia on the tissue concentrations of GABA in the brains of freshwater turtle, loggerhead sea turtle, and crucian carp. Crucian carp was exposed to 4.5 h of anoxia at 18 °C (group 1) or 17 days of anoxia at 8 °C (group 2). Freshwater turtle was exposed to 13 h of anoxia at 25 °C, and sea turtle to 4 h of anoxia at 25 °C. From Nilsson[114,118], Nilsson et al.[120,126]

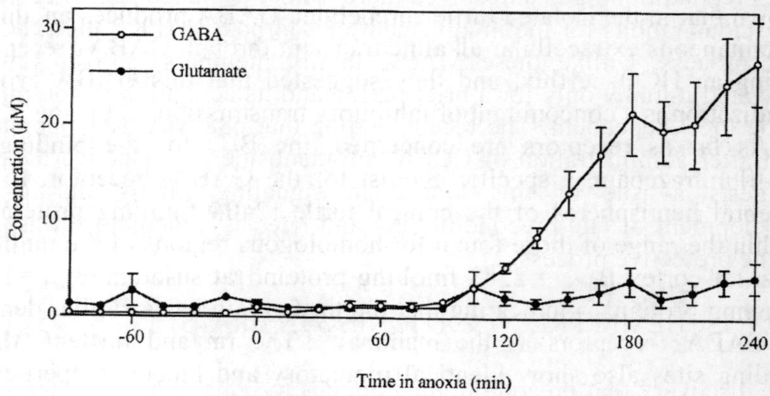

Fig. 7.13 Extracellular concentrations of GABA and glutamate measured by microdialysis in the brain (striatum) of freshwater turtles exposed to anoxia. Adapted from Nilsson and Lutz.[122]

However, anoxia provokes an up-regulation of $GABA_A$ receptors in the turtle brain. A rise in $GABA_A$ receptor number occurs within 2 h anoxia and there is a continued increase that reaches 129% of normoxic control levels within 24 h of anoxia.[99,154] This up-regulation in $GABA_A$ receptors may function to increase the effectiveness of the inhibitory action of GABA that is released during anoxia. Such an effect could be further enhanced by the decreasing extracellular pH that occurs during anoxia, since H^+ ions have a facilitatory action on $GABA_A$-receptor mediated responses.[73]

There are some interesting aspects of the responses of the mammalian GABAergic system to low oxygen levels. The effect of exogenous GABA has been found to be enhanced during anoxia in mammalian hippocampal slices, which is thought to be due to an enhancement of $GABA_A$ mediated responses by H^+.[84] However, in the gerbil hippocampus, 30-60 min ischemia results in a substantial down-regulation of $GABA_A$ receptors, possibly via receptor internalization.[1] This down-regulation could be a compensatory response to excessive GABA release. However, a short exposure to partial ischemia in the rat brain provides a long-lasting protection (7 days) against the convulsive agents bicuculine and pilocarpine, an effect the authors attributed to the induction or up-regulation of GABAergic mechanisms.[49,162] More directly, as little as 6 h exposure to moderate hypoxia (10% O_2) results in an increase in [^{3}H]flunitrazepam binding in the rat brain (21% increase in cortex).[128] But the time course of this process was quite different from that seen in the anoxic turtle since, with continued hypoxia, the B_{max} for the rat brain declined, returning to control levels within 24 h.[128] It is possible that after a period of adjustment, the moderately hypoxic rat is able to compensate for a lowered cerebral arterial oxygen content by increasing the cerebral blood flow and thereby restoring brain oxygen supply. By contrast, in the turtle brain, the increase in active $GABA_A$ receptors is part of an integrated metabolic response to bring about and consolidate a brain hypometabolic state.

One factor that could argue against a role of GABA in metabolic depression is receptor desensitization. $GABA_A$ receptors have been found to be desensitized and close their Cl^- channel within seconds or less when continuously stimulated with high concentrations of GABA.[78,103] However, how the receptor desensitization seen on the cellular level will affect GABAergic neurotransmission on a physiological level, with millions of GABA receptors involved, is unclear, since there is experimental evidence demonstrating that an increased GABAergic tonus can have very long lasting effects. For example, in mammals, an injection of GABA, or a drug that induces an increase in endogenous GABA, will cause a depression of the body temperature lasting several hours[154] Moreover, increased GABA receptor activation is thought to be involved in sleep as well as anesthesia,[73,195] processes that last far beyond the activation time for GABA receptor desensitization. It is of course also possible that GABA receptors in the brain of anoxia-tolerant vertebrates differ from mammals in their tendency to desensitize.

The brain plays an important role as an overall regulator of systemic physiological responses to changes in environmental factors. For anoxia-tolerant animals the neuronal changes brought about by a rise in GABA may not only depress brain metabolism, but may also control the decrease seen in systemic energy consumption during prolonged hypoxia. This is suggested by experiments on the metabolically depressed anoxic crucian carp, where treatment with anti-GABAergic drugs causes a considerable increase in metabolic rate, measured as ethanol excretion.[118] The same GABAergic drugs were found to be without effect on the normoxic metabolic rate (oxygen consumption).

5.2.2 Glycine and Taurine

Glycine is a well established inhibitory neurotransmitter in lower brain areas and the spinal chord of mammals, but in higher brain areas of mammals glycine is a potent allosteric activator of glutamate NDMA receptors[43,66] acting, therefore, as an excitatory factor. However, a maximal NMDA receptor activation seems to be reached already with 1 μM of glycine.[66] Because microdialysis measurements have indicated that the normal extracellular glycine level in the turtle brain is at least 3 μM[122] it is likely that any additional increase in glycine will have little or no excitatory effects.

Anoxia produces increased intracellular glycine concentrations in both the turtle[120] and crucian carp[114] brain. In the mammalian brain, by contrast, no increase in the tissue levels of glycine is seen during either anoxia or ischemia[37] which perhaps indicates an important difference between the anoxia-tolerant brain and the mammalian brain.

Taurine levels, which are fairly stable in the ischemic mammalian brain[88], increased in the anoxic turtle brain[120] but showed a modest decrease in the brain of the crucian carp.[114] The extracellular levels of glycine and taurine also increase during anoxia in the turtle brain (4 fold and 24 fold, respectively, after about 250 min in anoxia).[122]

Interestingly, the increases in extracellular concentrations of glycine and taurine occur at the same time as GABA is released in the anoxic turtle brain.[122] Whether or not taurine is a neurotransmitter is debated[55], but there is evidence that it is involved in the hyperpolarization of Purkinje cell neurons[130] and in protecting hypoxic neurons from Ca^{2+} influx.[107] Clearly, the possibilities that glycine or taurine mediate metabolic depression or otherwise protect the anoxic brain against anoxic damage deserves further experimental attention.

To conclude, it is likely that the release of inhibitory amino acids during anoxia plays an important role to maintain and even consolidate the depressed state of synaptic excitability that follows the initial metabolic down regulation (possibly mediated by adenosine).

5.3 Excitatory Amino Acids: Protection from Neurotoxic Effects

In contrast to GABA, tissue concentrations of the excitatory amino acid glutamate declines in the anoxic brains of both freshwater turtles[120] and crucian carp[114] (Fig. 7.14), while in the brains of anoxic mammals, the tissue levels of glutamate usually remain fairly constant during the short time they survive.[163] The decreases in glutamate may be related to the fact that the synthesis of neuronal glutamate is mainly aerobic since neuronal glutamate to a large extent appears to be derived from glutamine and a-ketoglutarate.[11]

During four hours of anoxia there is little or no increase in extracellular glutamate in the turtle brain (Fig. 8.12), which makes it strikingly different from the mammalian brain, where a massive release of glutamate occurs within minutes of anoxia and is a major cause of anoxic and ischemic brain damage. However, depolarization, produced by exposing the neurons to a high extracellular level of K^+ (delivered via the microdialysis probe), causes the anoxic turtle brain to release a mammalian-like surge of glutamate.[122] Significantly, turtle brains that have been depolarized do not recover evoked potential activity following reoxygenation[39,40], indicating that functional damage has been incurred. It would appear likely, therefore, that the turtle brain may be susceptible to the excitotoxic effects of a massive release of glutamate, well described for the mammalian brain, and that a primary defense is to avoid depolarization. Intriguingly, recent evidence indicates that glutamate continues to be released at controlled levels, in the anoxic turtle brain, but that extracellular levels are maintained through continued active uptake.[109a]

Both glutamate and aspartate evoke changes in the turtle cerebellar slice similar to those seen in the mammal, i.e. triggering spreading depression[86] and initiating increases in extracellular K^+ and a shift of extracellular Ca^{2+} into the cell.[147] In pyramidal cells in the dorsal cortex, excitation was observed when glutamate was applied to the so called "molecular layer of cortex, while a mixture of excitation and inhibition resulted from glutamate application to other layers of the cortex[82], the inhibitory effects possibly being the result of a secondary activation of GABAergic interneurons. Glutamate[82] and NMDA[152] also produce powerful activation of turtle nonpyramidal cells.

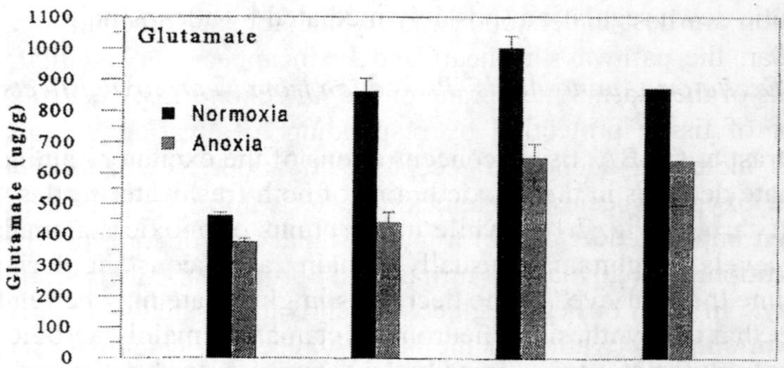

Fig. 7.14 Effects of anoxia on the tissue concentrations of glutamate in the brains of freshwater turtle, loggerhead sea turtle, and crucian carp. Crucian carp was exposed to 4.5 h of anoxia at 18 °C (group 1) or 17 days of anoxia at 8 °C (group 2). Freshwater turtle was exposed to 13 h of anoxia at 25 °C, and sea turtle to 4 h of anoxia at 25 °C. From Nilsson[114,118], Nilsson et al.[120,126]

The kinetics, pharmacology and regulation of the glutamate NMDA receptor complex is similar in rat and turtle brain[152] and no changes in NMDA receptor regulation were detected in the turtle forebrain following 2-4 h anoxia[153], indicating that the NMDA receptor is highly conserved in vertebrates. On the other hand, embryonic turtle cortical neurons, which also have NMDA and non-NMDA glutamate receptors that show characteristic glutamate-induced excitatory currents, are much more resistant to glutamate neurotoxicity than those of the mammal.[187] This resistance has been linked to the turtle brain having a much higher density of PBNO-opioid receptors compared to the rat.[191]

Interestingly, prolonged anoxia appears to modulate glutamate-gated Ca^{2+} channels in the turtle brain to produce a substantial reduction in glutamate mediated Ca^{2+} influx[10]. It has been suggested that a massive release of GABA, such as seen in the anoxic turtle brain, could cause the membrane to hyperpolarize and shift far from the range at which the NMDA receptors operate at maximal efficiency.[152] However, there is no evidence of such a shift in anoxic turtle brain slices[137], but on the other hand, we do not know if there is an anoxia induced GABA release in this *in vitro* preparation.

Adenosine may be involved in the modulation of NMDA receptor activity. Adenosine has been found to reduce the intracellular Ca^{2+} influx stimulated by NMDA in turtle cortical slices.[15] A study using cell-attached patch clamp[16] has indicated that both anoxia and adenosine reduce the open probability (P_{open}) of the NMDA receptor by as much as 68%, and that the adenosine invoked decrease in P_{open} could be prevented by application of the adenosine A1 receptor antagonist 8-phenyltheophylline. In mammals, adenosine is also known to reduce NMDA-receptor mediated excitatory potentials.[30]

5.4 Monoamines: the Oxygen Dependent Neurotransmitters

Since monoamines require molecular oxygen for their synthesis and parts of their degradation (see chapter 2) it is likely that anoxia strongly influences monoamine neurotransmitter systems and, perhaps, that these transmitters influence anoxia tolerance. Both norepinephrine and serotonin act as inhibitory neurotransmitters on most mammalian neurons.[161]

There appears to be a complete stop in serotonin metabolism in the brain of the anoxic crucian carp as judged from the complete absence of the main serotonin metabolite, 5-HIAA, after a few days of anoxia.[113] However brain levels of serotonin are remarkably well preserved over this period, with only a 24% decrease after 17 days of anoxia at 8 °C.[114] The turnover time for serotonin is much slower in the normoxic crucian carp brain (1 to 3 days) compared to the mammal (1-2 h)[115], indicating that the crucian carp is capable of a very high degree of recycling of serotonin molecules, thereby saving on the serotonin stores. Serotonin levels are similarly maintained for at least 13 h in the brain of anoxic turtles kept at room temperature.[120]

Brain concentrations of the other putative inhibitory monoamine norepinephrine are also preserved in the anoxic turtle and crucian carp brain.[115,120] Interestingly, in the turtle brain, the normoxic levels of serotonin, norepinephrine and epinephrine are about 4 to 10 times higher in turtles than in mammals, which suggests that the oxygen dependent inhibitory monoamines may be stockpiled in the turtle brain in order to last during prolonged anoxia.[120]

In contrast to the other monoamines, the dopamine content of turtle brain is similar to that seen in mammals, but the turtle dopamine levels are not maintained and slowly fall to about 50% after 13 h of anoxia at 25 °C.[120] In the mammal the uncontrolled dopamine release, which occurs even during comparatively mild hypoxia, may be a major cause of hypoxic damage.[46] For example, one effect of increased extracellular dopamine is the inhibition of Na^+/K^+ pump activity, and a resultant rise in intracellular Na^+[6]. By contrast in the turtle brain extracellular dopamine levels do not increase during prolonged anoxia.[109] But, most interestingly, it appears that dopamine continues to be released during anoxia, albeit at a lesser rate, and that the maintenance of extracellular dopamine levels is due to still active uptake mechanisms.[109] However, during the initial transition period from the normoxic to the anoxic state recent evidence suggests that dopamine release is inhibited via a K_{ATP} channel mediated process. In the mammal there appears to be two uptake processes for dopamine: one high affinity and one low affinity.[5] The high affinity is inhibited by extracellular K^+ and is therefore likely to be suppressed during anoxia when extracellular K^+ is elevated so that the functioning of low affinity uptake is crucial under these conditions.[5] Berndt et al.[5] found a 50% increase in the V max of the low affinity site in the rat striatum after 14 h hypobaric hypoxia (11% O_2) and suggested that this increased capacity in the dopamine carrier represented an adaptation to hypoxia by allowing a faster removal of dopamine.

5.5 *Summarizing the Roles of Neurotransmitters/Neuromodulators*

The pattern of release of neurotransmitters/neuromodulators in the anoxic turtle brain suggests the following model[98] (Fig. 7.15). Initially during hypoxia anaerobic glycolysis is stimulated to make up the deficiency in energy supply brought about by the fall in oxidative phosphorylation. However as anoxia is established the rate of ATP synthesis falls below that of ATP use and ATP concentrations start to fall. The early rise that is seen in extracellular adenosine is the result of this fall in ATP levels and a release of adenosine serves as a signal for the initiation of a coordinated set of compensatory responses which results in a drastically reduced energy demand coupled with an increase in glycolytic energy supply (glucose), The result is the recovery of cellular energy status . However, as brain ATP recovers, adenosine levels decline and its inhibitory effects are reduced. But during deep anoxia here is a sustained release of inhibitory neurotransmitters, which act to stabilize the anoxic brain by consolidating the hypometabolic state. By these means anoxic depolarization is avoided as well as the consequences of uncontrolled excitatory amino acid release. But for the brain the hypometabolic state is not simply a matter of suppressing energy demand , in order to recover the brain must maintain its complex functional integrity, one aspect of which is the preservation of neurotransmitter "tone" through the continued controlled-release of such, potentially lethal, neurotransmitters as glutamate and dopamine.

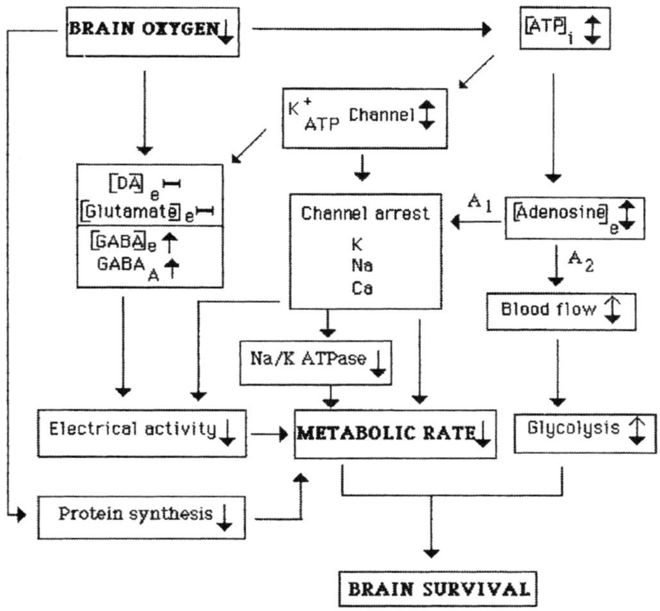

Fig. 7.15 Model of the role of the initial sequence of events involving ATP, adenosine and inhibitory neurotransmitters in the anoxia survival of the turtle brain. See text for details. Modified from Lutz and Nilsson.[98]

6. ION CHANNELS

6.1 Channel Arrest and Ion Loss

Because such a large proportion of cellular energy is used to maintain or restore ion gradients it was proposed by both Lutz et al.[101] and Hochachka[51], that a major reduction in ion flux during anoxia could be a key strategy to yield important energy savings. That this "channel arrest"[51] strategy does occur in the anoxic turtle brain was first indicated by the finding of Chih et al.[22] of a decreased rate of K^+ efflux in the anoxic brain compared to the normoxic brain after K^+ uptake had been blocked with ouabain (a Na^+/K^+ pump inhibitor).

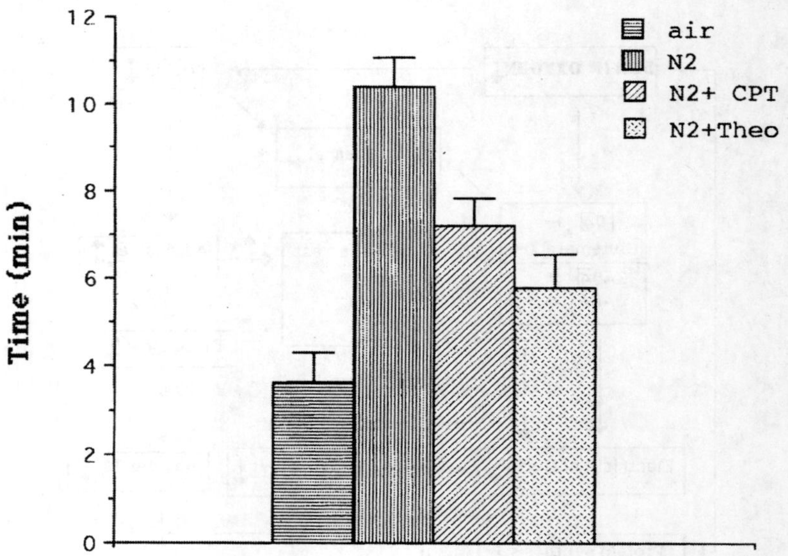

Fig. 7.16 Time for a 3mM increase in basal $[K^+]_o$. Following the application of ouabain, anoxic turtle brains superfused with the adenosine receptor blockers, theophylline or 8-CPT, show faster rises in $[K^+]_o$ (i.e. rates of K^+ efflux) compared to untreated anoxic brains. From Pek et al.[132]

There is a reduction in K^+ efflux to about 50% of normoxic values in 1 h and to about 70% at 2 h anoxia, after which no further decrease is seen.[133] The Ca channel NMDA receptor also down regulates over this period.[9] Adenosine appears to be a key in player in channel arrest. Increased resistance to K^+ leakage is partially mediated by adenosine receptors[132], and the ouabain induced K^+ efflux is significantly faster in anoxic turtle brains superfused with the specific adenosine A_1 blocker 8-CPT compared to untreated anoxic brains (Fig. 7.16). The anoxia induced down regulation of turtle brain NMDA channels is dependant on adenosine.

As discussed in Chapter 4, energy failure in the vertebrate brain (which would be produced by ouabain superfusion) results in a three phase change in extracellular K+ level , an initial slow increase (phase 1), until at a certain extracellular K^+ level (the inflection $[K^+]_o$), a rapid increase occurs (phase 2) until a plateau is reached (phase 3) corresponding to full depolarization. The inflection $[K^+]_o$ for normoxic ouabain treated turtle brain (8.22 mmol±0.87) is close to that found in adult rat brains.[132] However, this value is reduced by 50% in the anoxic turtle brain. Interestingly, in the rat cortex the inflection $[K^+]_o$ is strongly influenced by age or maturity, being about 24 mmol for the 4-12 day old neonate and declining to about 10 mmol for the adult.[47] The cause of this change in threshold is not known. In the rat brain stem, anoxic exposure results in a substantial depletion of intracellular K^+[64] so that, presumably, a greater increase in $[K^+]_o$ would be required to reach a fixed threshold transmembrane voltage, but such a loss of intracellular K^+ is almost certainly a pathological event and unlikely to be allowed by the anoxia adapted turtle brain. Alternatively, a decrease in K^+ conductance or an increase chloride conductance could mediate the depolarization effect of an increase in $[K^+]_o$.

In contrast to the turtle, in the crucian carp the rate of increase in $[K^+]_o$ after ouabain treatment is unaffected by anoxia[67], which may be taken as an indication of an unchanged K^+ permeability of the neuronal membrane. However, when interpreting the results of this type of experiment it is important to consider the factors responsible for the efflux of K^+ from neurons. Two are immediately apparent. Firstly, the number of K^+ channels that are open under resting conditions (i.e. between action potentials), and secondly, the electric activity which will lead to K^+ outflow through opened voltage dependent K^+ channels during action potentials. As we have discussed, one scenario in the anoxic brain of an anoxia-tolerant vertebrate is that electric activity is depressed due to the opening of K^+ channels gated by substances like adenosine and GABA. In such a situation, the flow of K^+ through voltage gated channels may be strongly suppressed at the same time as the number of opened receptor gated K^+ channels are increased. Therefore, the outflow of K^+ seen after ouabain treatment could remain the same in anoxia as in normoxia although K^+ is actually flowing through different types of channels.

Sick et al.[160] proposed that a net decrease in plasma membrane ion permeability, i.e. channel arrest, could be achieved via two distinct changes: (1) "Leakage arrest" whereby ion leakage is reduced when neurons are electrically inactive (i.e. not spiking), and (2) "Spike arrest" which may be brought about by inhibiting channel activity associated with action potentials or by reducing their activity by suppressing synaptic transmission.

Using intracellular microelectrodes, Perez-Pinzon et al.[137] found a decrease in input resistance of anoxic Purkinje cells in isolated turtle cerebellum. However, microelectrodes are not ideal for studying membrane resistance due to leakage currents between the electrode and the membrane. Here, the patch-clamp technique may be better suited. Still, Doll et al.[32], using patch clamp technique, reported no change in whole cell input resistance of turtle cortical pyramidal neurons during anoxia. However, more recently, Ghai and Buck[45] found that if EGTA is excluded from the patch pipette, a 20% reduction in whole cell membrane conductance is indeed seen when turtle cortical brain slices are exposed to anoxia. The fact that the Ca^{2+} chelator EGTA can block this effect indicates that intracellular Ca^{2+} is involved as an intracellular second messenger. Further experiments in the same study[45] indicated that adenosine, by activating A1 receptors and acting through a rise in intracellular Ca^{2+}, was responsible for initiating the reduced ion permeability of the turtle neurons during anoxia.

Unlike the arrest of leakage channels, "spike arrest" can cause substantial energy savings, and may be achieved by many mechanisms. One would be the release of inhibitory neurotransmitters, already discussed. Other possible mechanisms include changes in the properties or number of voltage-gated ion channels, as well as activation of various conductances that inhibit electric activity, i.e. primarily K^+ channels.

6.2 K^+ Channels

As mentioned in Chapter 4, the mammalian brain initially responds to anoxia with a short period of hyperpolarization. This important, albeit temporary, protective phase is associated with an increase in membrane conductance, which has been attributed, depending on the brain region, to the activation of ATP sensitive K^+ channels (K_{ATP} channels) or Ca^{2+} sensitive K^+ channels.[84]

The K_{ATP} channels may be pivotal, since by opening when cytosolic ATP concentrations decline and closing when cytosolic ATP levels are restored they directly respond to changes in the intracellular energy charge (see chapter 2). Moreover, K_{ATP} channels mediate vasodilation and therefore facilitate an increased blood supply when ATP levels fall. Evidence also points to the involvement of K_{ATP} channels in a wide variety of processes. For example, in the canine heart the K_{ATP} channel may have a synergistic interaction with the adenosine A1 receptor, both being involved in ischemic preconditioning[193] and there is evidence that the K_{ATP} channel may play a role in retarding excitatory neurotransmitter release (glutamate, dopamine) during the initial energy crisis of anoxia/ischemia.[174] Clearly it is of considerable interest to discover if such mechanisms operate in the brain of anoxia tolerant species.

In the crucian carp, glibenclamide, a specific blocker of mammalian K_{ATP} channels, was found to have no effect on the rate of K^+ outflow in brain (telencephalon) evoked by ouabain during anoxia.[67] However, the lack of effect of glibenclamide in the anoxic crucian carp brain may have phylogenetic background, rather than reflecting an adaptation aimed at enhancing anoxia tolerance. It may be that fishes in general lack glibenclamide binding K_{ATP} channels. In the rainbow trout brain, it was also found that the anoxia induced increase in the extracellular K^+ was unaffected by glibenclamide treatment[67] and [^{3}H] glibenclamide binding could not be detected in heart ventricles of lamprey, Chinook salmon and common carp.[106] However, the same study showed glibenclamide binding in the hearts of leopard frog (*Rana pipiens*), cane toad (*Bufo marinus*), rattlesnake (*Crotalus viridus*) and freshwater turtle (*Chrysemys picta*) at higher site densities than mammals. K_{ATP} channels that bind glibenclamide have been found in the clawed frog (*Xenopus*) nervous system[71], and glibenclamide binding sites in the turtle brain are about 10-30% that of the rat brain[61] a level that corresponds to the difference in metabolic intensity between their respective brains.[133]

In the turtle brain the addition of K_{ATP} channel blockers glibenclamide or 2,3-butanedione monoxide prevented the anoxia induced decrease in K^+ efflux during the first hour of anoxia, but the effect of these blockers had diminished at 2 h anoxia and was not seen after 4 h anoxia.[133] This pattern can be related to the initial fall and subsequent recovery in turtle brain ATP during early anoxia.[97] The mechanisms that link K_{ATP} channel activation and the initial reduction in K^+ efflux are unknown. In the mammal, membrane hyperpolarization is thought to be the primary protective effect of K_{ATP} channel activation, but as mentioned above there is much evidence that the turtle brain maintains membrane potential during prolonged anoxia. K_{ATP} channel activation acts to reduce dopamine efflux during early anoxia in the turtle striatum thereby maintaining of low extracellular dopamine levels[109], and similarly protects against early glutamate increase.[109a]

With respect to the overall protective function of K_{ATP} channels, it is important to recognize that if the animal is unable to recover ATP levels during anoxia the K_{ATP} channels may actually contribute to the anoxic catastrophe. A K^+ mediated hyperpolarization initiated by a falling ATP concentration can only serve as a hypoxia defense mechanism in the short term (min) and may be disastrous over longer anoxic periods. This is because the increase in K^+ conductance will result in an increased efflux of K^+ and shorten the time to reach the inflection $[K^+]_o$ which triggers the phase 2 loss of cell potential. This indeed is what happens in the mammalian brain after a few minutes of anoxia. Indeed, the K_{ATP} channels are thought to be a major route for the rapid efflux that occurs during anoxic depolarization in the mammalian cortex[189], but do not appear to have such importance in the depolarizing turtle brain.[133]

Another mechanism for eliciting K^+ mediated hyperpolarization is the activation of Ca^{2+} sensitive K^+ channels. There is good evidence that in the mammalian hippocampal neurons, where these channels are abundant, the hypoxia induced increase in cytosolic Ca^{2+} does indeed cause a major increase in K^+ conductance.[84] Unfortunately we have no information on distribution or activity of Ca^{2+} activated K^+ channels in either the turtle or crucian carp brain. However, brain slices of turtles maintain normal intracellular Ca^{2+} over prolonged anoxic conditions.[8]

In summary K_{ATP} channels play an important role during the down-regulation of metabolic demand that occurs as the turtle brain becomes anoxic. When the brain initially goes into energy imbalance the consequent fall in brain ATP acts as a signal which activates K_{ATP} channels , which in turn initiate processes that result in a reduction in K^+ efflux. This is not achieved via membrane hyperpolarization , but may involve other factors such as the inhibition of excitotoxic neurotransmitter release in the energy challenged brain, and the enhancement of the adenosine A1 receptor response. When a balance is achieved between (reduced) ATP demand and (glycolytic) ATP supply, brain ATP levels are restored and the K_{ATP} channels inactivated.

6.3 Na^+ Channels

Edwards et al.[36] found that the turtle synaptosome had a voltage gated Na^+ channel density 1/3 that of the rat synaptosome, but they concluded that this difference by itself was not an important intrinsic factor in the 100 fold difference in anoxia tolerance that exist between turtles and rats. Xia and Haddad[190] similarly found that for most brain regions, particularly the cerebellum, the Na^+ channel density of the turtle was much less than that of the rat, and reached the same conclusion. In a few areas, however, such as the medullary raphe nuclei, they found that the Na^+ channel density of the turtle was similar to or even higher than in the rat.

However, blocking the voltage dependent Na^+ channels with tetrodotoxin causes substantial decreases in energy consumption of rat and turtle synaptosomes (by 22% in turtle and by 19% in rat[36]) and a similar depression is seen in the *in situ* perfused rat brain (by about 20%).[192] Correspondingly, increasing Na^+ channel activation with veratridine increases energy consumption by about 20% in rat and turtle synaptosomes.[36,176] This indicates that modulation of Na^+ channel conductance could have a significant effect on metabolic cost.[36] Since voltage gated Na^+ channels are responsible for both the initiation and conduction of action potentials, modulation of their functional properties would be important determinants of brain electrical activity and thus energy consumption. The only direct evidence for this mechanism operating in the turtle comes from a study which found that 4 h of anoxia produced 42% decline in the density of voltage gated Na^+

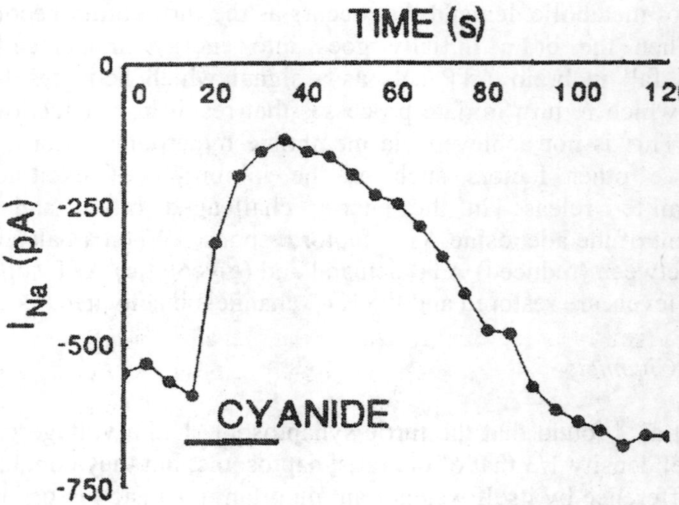

Fig. 7.17 Whole cell voltage clamp studies of human neocortical neurons show that both CN⁻ and anoxia produce a rapid inhibition of the fast inward Na^+ current (I_{Na}). From Cummings et al.[26]

channels in the isolated turtle cerebellum.[138] Such a mechanism may underlie the anoxia-induced increase in action potential thresholds in the isolated cerebellum[137] and cerebrocortex[16] discussed earlier in this chapter. Interestingly, whole cell voltage clamp studies of human neocortical neurons by Cummings et al.[26] show that CN^- and anoxia produce a rapid inhibition of the fast inward Na^+ current (I_{Na}) (Fig. 7.17). They suggested that oxygen deprivation increases the probability for Na^+ channels to be in the inactive state, and that this is an adaptive energy saving mechanism that delays the onset of the depolarizing phase.[26] An alternative (or perhaps complimentary) explanation is that the reduction in membrane Na^+ permeability retards the early influx of Na^+, that precedes, and may be a major causative factor of, cell depolarization.[192]

6.4 Ca^{2+} Channels

The opening of Ca^{2+} channels is the ubiquitous signal for neurotransmitter release and like Na^+, the inflow of Ca^{2+} into the neurons will cause depolarization. In addition, we have seen in chapter 4 that the inflow of Ca^{2+} into mammalian neurons is one of the principal causes of anoxic brain death. Thus, a down regulation of Ca^{2+} channels is likely to be an important mechanism for anoxic survival.

Indeed, that such a mechanism exists is suggested by experiments in freshwater turtles. In turtle cortical brain slices, exposure to anoxia reduces the rise in intracellular Ca^{2+} brought about by either blocking anaerobic glycolysis with iodoacetate (Fig. 7.18) or by depolarizing the cells with high concentrations of glutamate.[10] The reduction in Ca^{2+} influx was greater with increasing exposure times to anoxia, indicating a progressive decline in Ca^{2+} channel opening. Additional experiments showed that a similar reduction in Ca^{2+} inflow could be induced by an superfusing the slices with an unidentified plasma protein that is present in the blood of anoxic turtles (but not in normoxic individuals). If such a protein is produced also in the brain, then it could be a mediator of Ca^{2+} channel down regulation during anoxia in the turtle brain.[10] Bickler and coworkers[9] recently reported that NMDA receptors are silenced by at least three different mechanisms operating at different times during anoxia in the turtle brain: dephosphorylation which takes place over the first few minutes of anoxia, an intracellular Ca^{2+} dependent control mechanism which operates over the next several hours, and the removal of receptors from the cell membrane which takes days to weeks at 3 °C.

In the rat hippocampal slice depolarization-induced Ca^{2+} uptake is much slower during anoxia, possibly, like the turtle, indicating an anoxia induced inhibition of Ca^{2+} channels.[74] Mammals and turtles may share a similar mechanism for attenuating Ca^{2+} uptake in anoxia, although it only provides partial protection in the mammal, while it forms part of a full protection strategy in the turtle.

The possibility of changes in Ca^{2+} conductances in *Carassius* brain remains to be explored. However, it is of interest to note that endogenous antifreeze proteins that have the ability to block Ca^{2+} (and K^+) currents have been shown to occur in fish.[150] Thus, there is the possibility of a link between anoxia tolerance and freeze tolerance in fish.

7. pH AND CO_2

Animals that go into a hypometabolic state commonly experience enhanced CO_2 and H^+ levels, the result of hypoventilation and the switch to anaerobic glycolysis, and to some extent these internal changes may even contribute to depress metabolism in a coordinated way.[93]

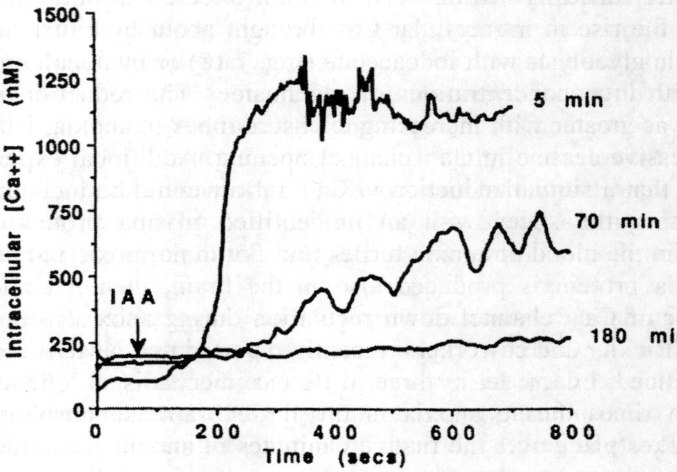

Fig. 7.18 Effects of iodoacetate superfusion (to block anaerobic glycolysis) on intracellular Ca^{2+} concentration in turtle brain cortical slices. Slices were preexposed to 5, 70, or 180 min of anoxia. These results indicate that anoxia results in a down regulation of Ca^{2+} conductivity. From Bickler.[8]

However, the tissue acidosis that occurs during anoxia or ischemia has been suggested to be a major perturbing factor for cell survival, since changes in intracellular pH have widespread metabolic effects.[164] It should be pointed out that anaerobic acidosis is brought about by the increase in lactate and from hydrolysis of ATP in excess of ATP formation.[53] Since ATP concentrations are maintained in the anoxic turtle brain, there will be no net production of H^+ from ATP hydrolysis. Also the hydrolysis of creatine phosphate at physiological pHs acts as a pH buffer by causing H^+ to be taken up on a mol per mol basis.[53] As the turtle brain becomes anoxic a reciprocal relationship is seen between the rise in tissue lactate and fall in phosphocreatine[97] indicating that the greatest changes in lactate are accompanied by the greatest uptake of

H^+ from phosphocreatine hydrolysis. The substantial increase in taurine content seen in the turtle brain during anoxia (50% increase within 1 h anoxia[120]), may also have a direct role in increasing the brains ability to buffer against acidosis[110]. The turtle brain therefore may have a greater degree of protection from changes in intracellular pH during anoxia compared to the mammalian brain. Nevertheless, the degree of acidosis experienced by the turtle during prolonged anoxia can be substantial. Six h of forced submergence resulted in the arterial pH falling from 7.7 to about 7.0 while the intracellular brain pH fell from about 7.45 to 6.9 in 4 h submergence.[94,185] Also, because CO_2 easily crosses the cell membrane, increases in blood PCO_2 levels are likely to be reflected in intracellular values. During long dives considerable amounts of CO_2 can build up, with arterial PCO_2 values as high as 100 Torr having been recorded.[62] The changes in the blood composition can be equally dramatic. For example in the freshwater turtle *Chrysemys picta* plasma lactate can rise to as much as 200 mM over 3 months submerged in cold (3 °C) anoxic water.[175] The increased lactate is accompanied by an enormous rise in plasma Ca^{2+} and Mg^{2+} concentrations, to as high as 50 mM Ca^{2+} and 15 mM Mg^{2+} which bind with the plasma lactate. Bone and shell are the most likely sources for Ca^{2+} and Mg^{2+} recruitment, probably accompanied by carbonate which would buffer H^+.[59]

The wide spread metabolic depressing effects of H^+ are well documented. Acid pH reduces glycolytic flux by, for example, promoting the phosphorylation of key regulatory enzymes (PFK, PK) to less active forms.[169] A reduction in pH causes a fall in glycolysis in both turtle and rat brain slices, but interestingly only in the rat brain is there a decrease in oxygen consumption.[18] In mammalian postsynaptic neurons a fall in pH reduces excitability, and an increase in the CO_2 level directly depresses the firing threshold of the mammalian hippocampal neurons.[2] Hypercapnia (20% CO_2 in the respiratory gas) has been found to depress olfactory bulb evoked potentials to a similar degree (25-30%) in turtles and rats, although the time course of the change was much slower in the turtle than in the rat.[41]

It is likely therefore that the large increases in CO_2 and H^+ that the anoxic turtle brain naturally experiences, are factors that not only have to be protected against but that may also be important in bringing about and maintaining the hypometabolic state.

The situation is quite different in the crucian carp and goldfish during anoxia. As mentioned above, instead of lactate, ethanol is here the main metabolic end product. Thus, in anoxia, these fishes have to cope with very modest lactate levels, less than 10 mM in the blood[179], when compared to turtles. This causes only minor changes in pH during anoxia. The formation of ethanol does not cause any net formation of H^+, but it involves a decarboxylation reaction that produces CO_2 and ethanol in equimolar amounts. However, CO_2 will not present a problem for these fishes since it is rapidly excreted by the gills.

8. FREE RADICALS

It is well known that free radicals are produced when ischemic mammalian tissue is reperfused. However, this is at least partly related to the energy depletion that mammalian cells suffer during anoxia/ischemia, which for example causes a rise in intracellular Ca^{2+}, breakdown of lipids, and degeneration of free-radical protecting mechanisms. Still, even if anoxia-tolerant animals maintain their ATP levels, the electron carriers of their mitochondria must be reduced and a subsequent reoxygenation could cause excessive oxyradical production. There is recent evidence for lipid peroxidation in goldfish brain (and liver). Thus, Lushchak et al.[92] found that the levels of conjugated dienes (indicators of lipid peroxidation) rose by 75% in brain during reoxygenation following anoxia. Also possible countermeasures against reoxygenation damage were detected: the activity of glutathione peroxidase was increased by 79% during anoxia and remained high during the subsequent reoxygenation, and the levels of glutathione were maintained both during anoxia and reoxygenation. The activities of other antioxidant enzymes, such as glutathione reductase, glutathione S-transferase, catalase and superoxide dismutase did not increase, but were maintained during anoxia and reoxygenation in goldfish brain. The freshwater turtle (*Trachemys scripta*) may be doing even better: there are experiments suggesting that lipid oxidation remains low, or is even suppressed, in freshwater turtle liver during reoxygenation following anoxia.[186]

Increasing attention is being given to the roles that ascorbic acid, a free radical scavenger, has in many important brain operations, particularly regarding its involvement in processes that directly relate to neurotransmitter function.[19,146] The rat brain maintains high extracellular concentrations of ascorbic acid (300-400 μM)[155,166], that can change during such conditions as depolarization[108], the release of excitatory amino acids[129] and hypoxia.[19] The turtle brain has greater concentrations of ascorbic acid levels, those in cortex levels being 2-3 times greater than in the mammalian cortex[149] and the isolated turtle cerebellum maintains reasonably constant extracellular ascorbate concentrations for many hours under normoxic conditions.[148] Even in the absence of glucose or oxygen, the turtle brain is able to maintain tissue ascorbate levels to as much as 80% of the control concentrations after 6 hours of anoxic incubation.[148] In guinea pig brain slices, by contrast, only 4% of the initial ascorbic acid content remains following 6 h incubation.[148] However, spreading depression, produces a large increase in the extracellular ascorbic acid concentration, through a decrease in the extracellular volume and a release of ascorbate from intracellular stores.[148] Rice and Cammack[148] speculate that the maintenance of high levels of ascorbic acid during anoxia could be an important factor to protect the turtle brain against reoxidative injury when normal respiration is resumed after anoxia.

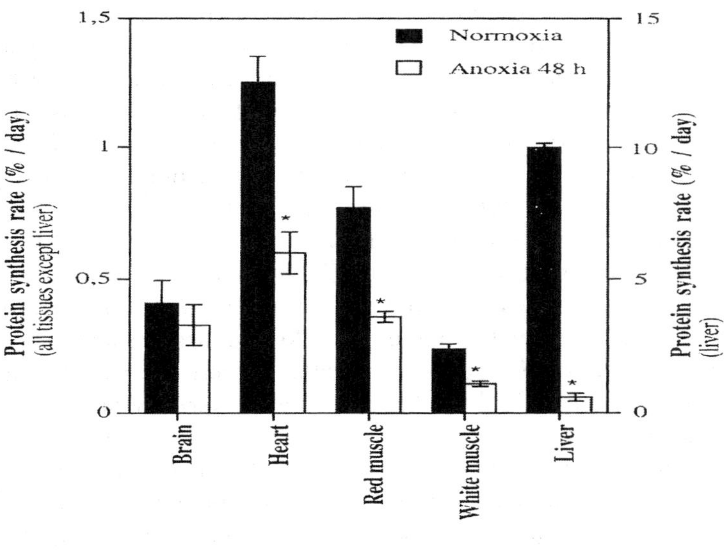

Fig. 7.19 Rates of protein synthesis in different tissues of crucian carp during normoxia and 48 h of anoxia at 9 °C. The rates were calculated from the incorporation of injected [^{3}H]phenylalanine to proteins in vivo. Data from Smith et al.[165]

9. PROTEIN SYNTHESIS

Down-regulation of protein synthesis has often been hypothesized to be a mechanism for metabolic depression. The first *in vivo* data on the normoxic and anoxic rates of protein synthesis in tissues of an anoxia tolerant vertebrate was published not that long ago.[165] In this study, normoxic and anoxic crucian carp were injected with [^{3}H]phenylalanine, and the incorporation of radioactivity into protein was measured in brain, liver, heart, and skeletal muscle (red as well as white). The results revealed a striking difference in response of the various tissues (Fig. 7.19).[165] While the anoxic liver had virtually turned off its protein synthesis (> 95% suppression), the anoxic brain showed no significant fall in the rate of protein synthesis. Heart and muscle showed intermediate responses, reducing their rate of protein synthesis by approximately 50%. These changes make functional sense. While protein synthesis may account for as much as 85% of the ATP consumption of the fish liver[131], it is only a small fraction of the brain energy consumption (about 3% according to Smith et al.[165] Thus, by turning off protein synthesis, the liver will more or less have solved its problem of matching anaerobic ATP production with ATP consumption. With regard to skeletal muscle, its fraction of the fish whole body mass is so large (about 50%) that any energy savings in this tissue will be of importance in the whole energy budget, and therefore important for prolonging anoxic survival (As we have mentioned earlier in this chapter, it is the total exhaustion of the liver glycogen store that finally limits anoxic survival in *Carassius*). However, for the brain, protein synthesis accounts for such a small fraction of the energy use that slowing it down will be of little use for defending brain ATP levels, and of no significance in the whole body energy budget. Another factor may also be of importance here. The absolute rate of protein synthesis in brain is only 4% of that of the liver (Fig. 7.19). Much of the protein synthesis of the liver is actually aimed for extra-hepatic use or needed in functions that are turned off during anoxia (such as food digestion). By contrast, the proteins synthesized by the brain are largely for local use, and it is likely that in order to maintain brain function, protein synthesis has also to be maintained.

Nevertheless, in the turtle, preliminary results indicate a profound down regulation of protein synthesis in its brain during anoxia. Moreover, there is evidence that anoxia differentially affects expression of proteins in the optic system of turtles (*Trachemys scripta*), some proteins being down regulated, probably to conserve energy, others being up regulated, probably to provide protective mechanisms (McFadden, Lutz & Perry, unpublished). With regard to turtle liver, there is *in vitro* data[85] on cultured freshwater turtle (*Chrysemys picta*) hepatocytes showing that anoxia produced a 92% suppression of protein synthesis.

10. BRAIN SWELLING
The Reason for having a Large Brain Cavity?

When the mammalian brain becomes anoxic the uncontrolled flow of ions into the anoxic cells is accompanied by osmotic driven water and the brain swells. But the close match between brain tissue volume and the containing skeletal brain cavity does not allow for any expansion in tissue volume and swelling leads to increased intracranial pressure. When this pressure rises above systolic blood pressure the outcome is global brain ischemia, a point of no return. A recent study using vivo nuclear magnetic resonance imaging (MRI) monitored changes in brain volume in two related fish species exposed to anoxia, the anoxia tolerant crucian carp and the hypoxia accommodating common carp (*Cyprinus carpio*).[178] The crucian carp maintained brain volume, even after 24 h of anoxia, with no signs of edema. By contrast in the common carp brain volume increased by 6% after 2 h in anoxia and a further 10% increase in brain volume was observed after 100 min of subsequent recovery (Fig. 7.20). In addition, clear signs of cellular edema was seen in the common carp brain. For a mammal the pressure increase corresponding to such an increase in brain volume would mean circulatory stop and death. An interesting feature of the scull of many ectothermic vertebrates, including these two species of carp, is that the cranial cavity volume is much greater than that of the brain. Much of the scull is filled with fluid, and a jelly like primitive meninx. The results of this study prompted the authors[178] to hypothesize that the oversized brain cavity seen in many ectothermic vertebrates, serves to allow some degree of brain swelling without an increase in intracranial pressure. This would reduce the risk of irreversible brain ischemia in situations that cause brain energy deficiency, such as hypoxia and anoxia.

11. FRESHWATER TURTLES VERSUS CARASSIUS
Glycolytic Depression versus Glycolytic Activation

The observant reader will have noticed that there are some striking differences between the anoxia responses of the fish species, genus *Carassius*, and those of the freshwater turtles. Since there are several hundred million years of evolution between turtles, *Carassius* and a common ancestor, anoxia tolerance will obviously have evolved separately in these two animal groups.

The most important behavioral difference is that the anoxic turtles become completely unresponsive to external stimuli and go into a comatose state after about an hour. By contrast anoxic crucian carp and goldfish still swim around spontaneously, although at a reduced rate. Obviously, these differences are reflections of divergent adaptive responses in the brains of these animals.

In the turtle, brain blood flow is only temporarily increased during the first hour or so of anoxia, indicating that glycolytic activation here plays the role of an initial emergency response, before metabolic depression has become fully operational. This scenario also fits well with biochemical studies on freshwater turtles, which suggests that after the first hour of anoxia, key enzymes associated with glycolysis are turned into less active forms.[80] Phosphofructokinase, in particular, shows increased inhibition by ATP and citrate, and a decreased sensitivity to its activator AMP, and glycogen phosphorylase-a activity is reduced by 70%, indicating a decrease in the breakdown of glycogen to glucose.[80] By producing less active forms of the enzymes, the modification of key glycolytic enzymes by phosphorylation is possibly necessary for matching the rate of glycolysis in the turtle brain with the depressed rate of ATP use.[170] The fall in pH that accompanies prolonged anoxia in turtles can also act to depress enzyme activities, especially those associated with glycolysis.[170]

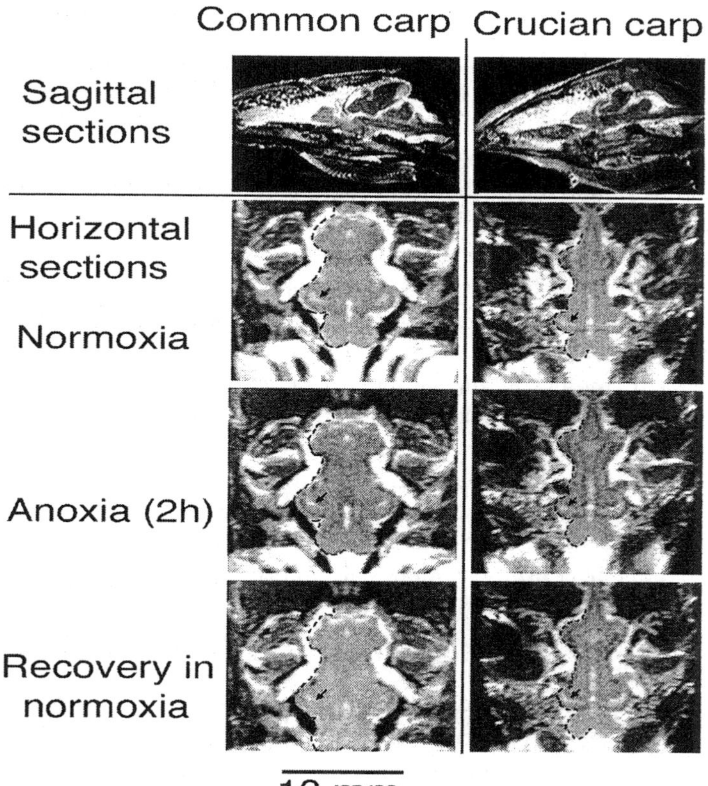

Fig. 7.20 Brain volume of anoxic carps studied by in vivo nuclear magnetic resonance imaging (MRI). This study revealed brain swelling in common carp exposed to anoxia. In the 1st row, sagittal sections of the heads of common carp and crucian carp are shown. Note that the brain (gray) is surrounded by an oversized brain cavity (white) The horizontal sections (rostral part of brain downwards) show the brains in normoxia (2nd row), after 2 hours of anoxia (3rd row), and after 100 min of subsequent recovery (4th row). During 2 hours of anoxia, the common carp brain (left column) swells by 6%, a process that continues into the recovery phase (10% swelling). A noticeable sign of the swelling is the compression of the ventricles (the whitish areas within the gray brain tissue, marked by arrows). A broken line marks the outline of the left half of the brains. From Van der Linden et al.[178]

In contrast to the turtle, experiments with crucian carp brain slices suggest a sustained 2.4 fold increase in the rate of glycolysis during anoxia, as judged from microcalorimetric measurements of heat output and lactate production.[68] Moreover, even after 24 h of anoxia (7 °C), the goldfish brain displays a 3.5 fold increase in the level of fructose 2,6-bisphosphate, a phosphofructokinase activator.[168] This was combined with an increase of the active (phosphorylated) a-form of glycogen phosphorylase in brain, while the active form of this enzyme was suppressed in all other tissues examined.[168] As mentioned, there is also the information that protein synthesis is not reduced in the anoxic crucian carp brain[165] while preliminary results suggest a profound down regulation of protein synthesis in the anoxic turtle brain.

Another difference is the lack of evidence for channel arrest in the brain of anoxic crucian carp.[67] For turtles, evidence for a decrease in neuronal ion permeability during anoxia is now accumulating. For example, while Chih et al.[22] found that anoxia causes a 50% reduction in the initial rate of extracellular K^+ increase after ouabain treatment, Johansson & Nilsson[67] found that the rate of ouabain induced K^+ efflux in the crucian carp brain was identical in normoxia and anoxia.

Also the release of GABA in the anoxic crucian carp brain appears to be much smaller (a doubling or so after 6 h) than that of the anoxic turtle brain (more than 50 fold). This difference fits well with the striking difference in activity level between anoxic *Carassius* and anoxic turtles – active versus comatose. Possibly, one may regard GABA as an endogenous anaesthetic in the anoxic situation (indeed, many anaesthetics appear to work through GABA receptors[140], where the turtle relies on the deepest possible anaesthesia to save energy and protect the brain. By contrast, the crucian carp relies on a more subtle and dynamic down-regulation involving a limited regulated release of GABA.

The contrasting strategies of the turtle and *Carassius* represent two divergent adaptations for anoxic brain survival[100] in one (turtle) there is a maximal shut down to deep metabolic depression, in the other (*Carassius*) glycolysis is enhanced to allow co-ordinated activity during anoxia. The first strategy gives a comatose brain while the other results in a brain that is still "functioning". The key adaptation that allows a continued high level of glycolysis in *Carassius* is ethanol production. Hereby these species avoid the self-pollution of lactate producing glycolysis.

The selection pressure on turtles to suppress their metabolism must be very strong since they have to cope with enormous lactate loads. Therefore, any increase in metabolic rate during anoxia would shorten the anoxia endurance of the overwintering turtle. The capacity to produce ethanol then, liberated the crucian carp from the path taken by anaerobic lactate producers to a more and more suppressed metabolism. Ethanol production may have arisen from a single mutation that made the pyruvate dehydrogenase complex start leaking acetaldehyde.

However, one significant drawback with the ethanol strategy is that ethanol excretion represents a loss of an energy rich substrate. In fact, in stead of suffering from the build up of an anaerobic end product, it is the total exhaustion of the liver glycogen store that is the limiting factor for anoxic survival in *Carassius.*[116] Thus, the time that *Carassius* can endure anoxia is linearly related to its degree of metabolic depression. It is likely, therefore, that when faced with anoxia *Carassius* rapidly down regulates their rate of metabolism to a degree that is still compatible with maintaining activity and responsiveness to relevant external stimuli, but at the same time substantially lowers their rate of glycogen store depletion.

For *Carassius* the benefits gained from being able to be active in anoxia could be well worth this fuel loss. The amount of dissolved oxygen can vary a lot in a freshwater habitat and for a water breather, being active in anoxia allows you to come to the oxygen rather than to have to wait for oxygen to come to you. In the brain of *Carassius*, there may be a certain selectivity in which functions are down regulated and which are maintained. A modality that is likely to be maintained is the ability to sense the external oxygen tension, which in fish is done through receptor cells on the gills.[17] If the fish can monitor the external oxygen level, then it will probably be able to move towards higher oxygen levels. However, one function that appears to be down regulated is vision. As mentioned, anoxia rapidly causes complete suppression of the potentials that light directed against the eye evokes in the retina (electroretinogram) and the optic tectum.[69] Similarly, sound-evoked electric activity in the auditory system is depressed in anoxic goldfish.[38,172] Vision and hearing are probably of little importance for the crucian carp during anoxia, and these senses can therefore be temporarily sacrificed in order to save energy.

12. CONCLUSIONS

We can conclude that a key to anoxic survival is an ability to continue to match ATP consumption with ATP production in the absence of oxygen, thereby avoiding energy failure and consequent neuronal depolarization. In the freshwater turtle, this is accomplished by a rapid temporary increase in the rate of anaerobic glycolysis followed by a long term depression of energy needs which sends the turtle into a completely inactive (comatose) state. The crucian carp and the goldfish also lower their metabolic rate substantially, but to a lesser extent in order to allow physical activity during anoxia. This ability is probably closely related to the capacity of *Carassius* to produce ethanol, rather than lactate, as the major end product of anaerobic ATP production.

Likely mechanisms for the metabolic depression in brain include the suppression of neurotransmission caused by a modulation of ion channel conductances and the release of inhibitory neuromodulators like adenosine and inhibitory neurotransmitters such as GABA. At present, the most likely scenario is that of a battery of mechanisms working in concert to control and induce metabolic depression in anoxia tolerant animals.

The glycolytic ATP production, the other essential part of the anoxic survival strategy used by these animals, is supported by extra large glucose (glycogen) stores. In addition, there is a redistribution of blood flow to direct the glucose available to anoxia sensitive organs like the brain, changes in which adenosine appears to play another key role.

From the present state of knowledge, it appears that the physiological adaptations that allow anoxia-tolerant vertebrates to survive anoxia are mainly quantitative (such as larger glycogen stores and a more profound metabolic depression), and only to a lesser extent qualitative (e.g. ethanol production in *Carassius*). In other words, they seem to largely utilize mechanisms that to some degree are also present in anoxia-sensitive vertebrates like mammals. By up-regulating glycolysis a few times and down regulating ATP consumption by some 50-90% during anoxia, they avoid anoxic depolarization, and by defeating this critical threshold they achieve a 1,000 fold increase in anoxic survival time compared to “normal” vertebrates.

13. REFERENCES

1. Alicke B, Schwartz-Bloom RD. Rapid down-regulation of $GABA_A$ receptors in the gebril hippocampus following transient cerebral ischemia. J Neurochem 1995; 65:2808-2811.

2. Balestrino M, Somjen GG. Concentration of carbon dioxide, interstitial pH, and synaptic transmission in hippocampal formation of the rat. J Physiol 1988; 396:247-266.
3. Bernier NJ, Harris J, Lessard J, Randall D. Adenosine receptor blockade and hypoxia-tolerance in rainbow trout and Pacific hagfish. In: Effects on anaerobic metabolism. J Exp Biol 1996; 199:485-495.
4. Bernier NJ, Fuentes J, Randall DJ. Adenosine receptor blockade and hypoxia-tolerance in rainbow trout and Pacific hagfish. II. Effects on plasma cathecholamines and erythrocytes. J Exp Biol 1996; 199:497-507.
5. Berndt C, Henke W, Gross J. Hypoxia induces different responses of striatal high and low affinity dopamine uptake sites. Mol and Chem Neuropath 1993; 18:179-187.
6. Bertorello AM, Hopfield JF, Apenia A, Greengard P. Inhibition by dopamine of (Na^+-K^+) ATPase activity through D_1 and D_2 dopamine receptor synergisms. Nature 1990; 347:386-388.
7. Bickler PE. Effects of temperature and anoxia on regional cerebral blood flow in turtles. Am J Physiol 1992; 262:R538-R541
8. Bickler PE. Cerebral anoxia tolerance in turtles: regulation of intracellular calcium and pH. Am J Physiol 1992; 263:R1298-R1302.
9. Bickler,PE, Donohoe,PH, Buck LT. Hypoxia-induced silencing of NMDA receptors in turtle neurons. J Neurosci 2000, 20:3522–3528
10. Bickler PE, Gallego SM. Inhibition of brain calcium channels by plasma proteins from anoxic turtles. Am J Physiol 1993; 265:R277-R281.
11. Bradford HF. Chemical Neurobiology. New York: W.H. Freeman, 1986.
12. Branco LG, Steiner AA, Tattersall GJ, Wood SC. Role of adenosine in the hypoxia-induced hypothermia of toads. Am J Physiol 2000; 279:R196-201.
13. Brooks SJP, Storey KB. Anoxic brain function: molecular mechanisms of metabolic depression. FEBS Lett 1988; 232:214-216.
14. Brooks SPJ, Storey KB. Control of metabolic rate by multienzyme complexes: Is glycolytic rate in hypoxia and anoxia regulated by enzyme complex formation? In: Surviving Hypoxia: Mechanisms of Control and Adaptation. P. W. Hochachka, P. L. Lutz, T. Sick, M. Rosenthal and G.van den Thilart (Eds). Boca Raton, FL: CRC Press, 1993: 281-294.
15. Buck LT, Bickler PE. Role of adenosine in NMDA receptor modulation in the cerebral cortex of an anoxia-tolerant turtle (*Chrysemys picta belli*). J Exp Biol 1995; 198:1621-1628.
16. Buck LT, Bickler PE. Adenosine and anoxia reduce N-methyl-D-aspartate receptor open probability in turtle cerebrocortex. J Exp Biol 1998; 201:289-297.
17. Burleson ML, Milsom WK. Sensory receptors in the first gill arch of rainbow trout. Respir Physiol 1993; 93:97-110.
18. Caliguri MA, Robin E. Prolonged diving and recovery in the freshwater turtle, *Pseudemys scripta* -IV. The effects of profound acidosis on O_2 consumption in turtle vs rat (mammalian) brain and heart slices. Comp Biochem Physiol 1985; 81A:603-605.
19. Cammack J, Ghasemzadeh B, Adams RN. Electrochemical monitoring of brain ascorbic acid changes associated with hypoxia, spreading depression and seizure activity. Neurochem Res 1992; 17:23-27.
20. Chen JC, Chessler M. A bicarbonate dependent increase in extracellular pH mediated by $GABA_A$ receptors in turtle cerebellum. Neurosci Lett 1990; 116:130-135.
21. Chih CP, Feng ZC, Rosenthal M, Lutz PL, Sick TJ. Energy metabolism, ion homeostasis, and evoked potentials in anoxic turtle brain. Am J Physiol 1989; 257:R854-R860.
22. Chih CP, Rosenthal M, Sick TJ. Ion leakage is reduced during anoxia in turtle brain: a potential survival strategy. Am J Physiol 1989; 255:R338-R343.
23. Clark VM, Miller AT. Studies on anaerobic metabolism in the freshwater turtle (*Pseudemys scripta elegans*). Comp Biochem Physiol 1973; 44A:55-62.
24. Collis MG. The vasodilator role of adenosine. Pharmac Ther 1989; 41:143-162.
25. Crawshaw LI, Wollmuth LP, O'Connor CS. Intracranial ethanol and ambient anoxia elicit selection of cooler water by goldfish. Am J Physiol 1989; 256:R133-R137.

26. Cummings TR, Jiang C, Haddad GG. Human neocortical excitability is decreased during anoxia via sodium channel modulation. J Clin Invest 1993; 91:608-615.
27. Davies DG. Distribution of systemic blood flow during anoxia in the turtle, *Chrysemys scripta*. Respir Physiol 1989; 78:383-390.
28. Davies DG. Chemical regulation of cerebral blood flow in turtles. Am J Physiol 1991; 260:R382-R384.
29. Davies DG. Temperature and cerebral blood flow regulation in the freshwater turtle, Pseudemys scripta. Respir Physiol 1994; 95:329-335.
30. De Mendonca A, Ribeiro JA. Adenosine inhibits the NMDA receptor-mediated excitatory postsynaptic potential in the hippocampus. Brain Res 1993; 606:351-356.
31. Doll CJ. Anoxic CNS membrane: mechanisms of collapse and stabilization. In: Surviving Hypoxia: Mechanisms of Control and Adaptation. P. W. Hochachka, P. L. Lutz, T. Sick, M. Rosenthal and G.van den Thilart (Eds). Boca Raton, FL: CRC Press, 1993: 390-400.
32. Doll CJ, Hochachka PW, Reiner PB. Channel arrest: implications from membrane resistance in turtle neurons. Am J Physiol 1991; 261:R1321-R1324.
33. Doll CJ, Hochachka PW, Hand SC. A microcalorimetric study of turtle cortical slices: insight into brain metabolic depression. J Exp Biol 1994; 191:141-153.
33. Duncan JA, Storey KB. Subcellular enzyme binding and the regulation of glycolysis in anoxic turtle brain. Am J Physiol 1991; 262:R517-R523.
34. Dunn JF, Hochachka PW. Metabolic responses of trout (*Salmo gairdneri*) to acute environmental hypoxia. J Exp Biol 1986; 123:229-242.
36. Edwards R, Lutz PL, Baden D. Relationship between energy expenditure and ion channel function in the rat and turtle brain. Am J Physiol 1989; 255:R1345-R1359.
37. Erecinska M, Nelson D, Wilson DF, Silver IA. Neurotransmitter amino acids in the CNS. I. Regional changes in amino acid levels in rat brain during ischemia and reperfusion. Brain Res 1984; 304:9-22.
38. Fay RR, Ream TJ. The effects of temperature change and transient hypoxia on auditory nerve fibre response in the goldfish (*Carassius auratus*). Hearing Res 1992; 58:9-18.
39. Feng ZC, Sick TJ, Rosenthal M. Orthodromic field potentials and recurrent inhibition during anoxia in turtle brain. Am J Physiol 1988; 255:R484-R491.
40. Feng ZC, Rosenthal M, Sick TJ. Suppression of evoked potentials with continued ion transport during anoxia in turtle brain. Am J Physiol 1988; 255:R478-R484.
41. Feng ZC, Sick TJ, Rosenthal M. Extracellular pH and supression of electrical activity during anoxia in turtle and rat brain. Am J Physiol 1990; 258:R205-R210.
42. Fernandes JA, Lutz PL, Tannenbaum A, Todorov AT, Liebovitch L, Vertes R. Electroencephalogram activity in the anoxic turtle brain. Am J Physiol 1997; 273:R 911-R919..
43. Foster AC, Kemp JA. Glycine maintains excitement. Nature 1989; 338:377-378.
44. Fujiwara N, Higashi H, Shimoji K, Yoshimura M. Effects of hypoxia on rat hippocampal neurones in vitro. J Physiol 1987; 384:131-137.
45. Ghai HS, Buck LT. Acute reduction in whole cell conductance in anoxic turtle brain. Am J Physiol 1999; 277:R887-R893.
46. Globus MY, Busto R, Dietrich WD, Martinez E, Valdes I, Ginsberg MD. Effects of ischemia on the in vivo release of striatal dopamine, glutamate and γ-aminobutyric acid studied by intracerebral microdialysis. J Neurochem 1988; 51:1455-1464.
47. Hansen A.J. Extracellular potassium concentration in juvenile and adult rat brain cortex during anoxia. Acta Physiol Scand 1977; 99:412-420.
48. Hawkins RA, Mans AM. Intermediary metabolism of carbohydrates and other fuels. In: Lajtha A, ed. Handbook of Neurochemistry 2nd ed. Metabolism of the Nervous System., New York: Plenum Press,1983: 259-294.
49. Heim C, Bortolotto ZA, Cavalheiro EA, Sontag KH. Transient occlusion of rat carotid arteries decreases susceptibility to pilocarpine seizures. Brain Res 1991; 544:253-259.
50. Hitzig BM, Kneussl MP, Shih V, Brandstetter RD, Kazemi H. Brain amino acid concentrations during diving and acid-base stress in turtles. J App Physiol 1985; 58:1751-1754.

51. Hochachka PW. Defence strategies against hypoxia and hypothermia. Science 1986; 231:234-241.
52. Hochachka PW, Guppy M. Metabolic arrest and the control of biological time. Cambridge, Mass: Harvard Univ Press,1987.
53. Hochachka PW, Mommsen TP. Protons and anaerobiosis. Science 1983; 219:1391-1397.
54. Hochachka PW, Somero GN. Biochemical Adaptation. Princeton: Princeton University Press,1984.
55. Huxtable RH. Physiological actions of taurine. Physiol Rev 1992; 72:101-163.
56. Hylland P, Nilsson GE. Evidence that Acetylcholine mediates increased cerebral blood flow velocity in crucian carp through a nitric oxide-dependent mechanism. J Cerebral Blood Flow Metab 1995; 15:519-524.
57. Hylland P, Nilsson GE, Lutz PL. Time course of anoxia induced increase in cerebral blood flow rate in turtles: evidence for a role of adenosine. J Cerebral Blood Flow Metab 1994; 14:877-881
58. Hylland P, Nilsson GE, Lutz PL. Role of nitric oxide in the elevation of cerebral blood flow induced by acetylcholine in the turtle. J Cerebral Blood Flow Metab 1996; 16:290-295.
59. Hylland P, Nilsson, GE. Extracellular levels of amino acid neurotransmitters during anoxia and forced energy deficiency in crucian carp brain. Brain Res. 823: 49-58, 1999
60.. Hyvärinen H, Holopainen IJ, Piironen J. Anaerobic wintering of crucian carp (*Carassius carassius L.*) I. Annual dynamics of glycogen reserves in nature. Comp Biochem Physiol 1985; 82A:797-803.
61. Jackson DC. Metabolic depression and oxygen depletion in the diving turtle. J Appl Physiol 1968; 24:503-509.
62. Jackson DC. Acid-base regulation of reptiles. In: Heisler N, ed. Acid-base regulation in animals. Amsterdam: Elsevier Biomedical Press,1986: 235-263.
63. Jackson DC, Heisler N. Plasma ion balance of submerged anoxic turtles at $3^{o}C$: the role of calcium lactate formation. Respir Physiol 1982; 49:159-174.
64. Jiang C, Haddad GG. Effect of anoxia on intracellular and extracellular potassium activity in hypoglossal neurons in vitro. J Neurophysiol 1991; 66:103-111.
65. Jiang C, Xia Y, Haddad GG. Role of ATP-sensitive K+ channels during anoxia: major differences between rat (newborn and adult) and turtle neurons. J Physiol (Lond) 1992; 448:599-612.
66. Johnson JW, Ascher P. Glycine potentiates the NMDA response in cultured mouse brain neurons. Nature 1987; 325:529-531.
67. Johansson D, Nilsson GE. Roles of energy status, K_{ATP} channels and channel arrest in fish brain K^{+} gradient dissipation during anoxia. J Exp Biol 1995; 198:2575-2580.
68. Johansson D, Nilsson GE, Törnblom E. Effects of anoxia on energy metabolism in crucian carp brain slices studied with microcalorimetry. J Exp Biol 1995; 198:853-859.
69. Johansson, D., Nilsson, G. E., Døving, K. B. Anoxic depression of light-evoked potentials in retina and optic tectum of crucian carp. Neurosci. Lett. 1997; 237: 73-76.
70. Johnston IA, Bernard LM. Utilization of the ethanol pathway in carp following exposure to anoxia. J Exp Biol 1983; 104:73-78.
71. Jonas P, Koh DS, Kampe K, Hermsteiner M, Vogel W. ATP-sensitive and Ca-activated K channels in vertebrate axons: Novel links between metabolism and excitability. Pflugers Arch 1991; 418:68-73.
72. Juhász G, Kékesi KA, Emri Z, Ujszaszi J, Krogsgaard-Larsen P, Schousboe A. Sleep promoting effect of a putative glial g-aminobutyric acid uptake blocker applied in the thalamus of cats. Eur J Pharmacol 1991; 209:131-133.
73. Kaila K. Ionic basis of $GABA_A$ receptor channel function in the nervous system. Prog Neurobiol 1994; 42:489-537.
74. Kass IS, Abramowicz AE, Cottrell JE, Amorim P, Chambers G. Anoxia reduces depolarization induced calcium uptake in the rat hippocampal slice. Brain Res 1994; 633:262-266.

75. Kardos J, Cash DJ. Transmembrane $^{36}Cl^-$ flux measurements and desensitization of the γ-aminobutyric $acid_A$ receptor. J Neurochem 1990; 55:1095-1099.
76. Kauppinen RA, Nicholls DG. Failure to maintain glycolysis in anoxic nerve terminals. J Neurochem 1989; 47:1864-1869.
77. Keiver KM, Hochachka PW. Catecholamine stimulation of hepatic glycogenolysis during anoxia in the turtle *Chrysemys scripta*. Am J Physiol 1991; 261:R1341-R1345.
78. Keiver KM, Weinberg J, Hochachka PW. The effect of anoxic submergence and recovery on circulating levels of catecholamines and corticosterone in the turtle, *Chrysemys picta*. Gen Comp Endocrinol 1992; 85:308-315.
79. Keiver KM, Weinberg J, Hochachka PW. Roles of catecholamines and corticosterone during anoxia and recovery at 5°C in turtles. Am J Physiol 1992b; 263:R770-R774.
80. Kelly DA, Storey KB. Organ-specific control of glycolysis in anoxic turtles. Am J Physiol 1988; 255:R774-R779.
81. Keung EC, Li Q. Lactate activates ATP-sensitive potassium channels in guinea pig ventricular myocytes. J Clin Invest 1991; 88:1772-1777.
82. Kriegstein AR, Connors BW. Cellular physiology of the turtle visual cortex: synaptic properties and intrinsic circuity. J Neurosci 1986; 6:178-191.
83. Krilowicz BL, Glotzbach SF, Heller HC. Neuronal activity during sleep and complete bouts of hibernation. Am J Physiol 1988; 24:1008-1019.
84. Krnjevic´ K. Membrane current activation during hypoxia in hippocampal neurones. In: Surviving Hypoxia: Mechanisms of Control and Adaptation. P. W. Hochachka, P. L. Lutz, T. Sick, M. Rosenthal and G.van den Thilart (Eds). Boca Raton, FL: CRC Press, 1993: 365-388.
85. Land SC, Buck LT, Hochachka PW. Response of protein synthesis to anoxia and recovery in anoxia-tolerant hepatocytes. Am J Physiol 1993; 256:R41-R48.
86. Lauritzen M, Rice ME, Okada Y, Nicholson C. Quisqualate, kainate, and NMDA can initiate spreading depression in the turtle cerebellum. Brain Res 1988; 475:317-327.
87. Lee KS, Lowenkopf T. Endogenous adenosine delays the onset of hypoxic depolarization in the rat hippocampus in vitro via an action at A_1 receptors. Brain Res 1993; 609:313-315.
88. Lehmann A, Hagberg H, Andine P, Ellren K. Taurine and neuronal resistance to hypoxia. FEBS Lett 1988; 233:437-438.
89. Leone AM, Palmer RM, Knowles RG, Francis PL, Ashton DS, Moncada S. Constitutive and inducible nitric oxide synthase incorporate molecular oxygen into both nitric oxide and citrulline. J Biol Chem 1991; 266:23790-23795.
90. Lipton P, Whittingham TS. Energy metabolism and brain slice function. In: Dingledine R, ed. Brain Slices. New York: Plenum Press, 1984: 113-153.
91. Lowry OH, Passonneau JV, Hasselberger FX. Effects of ischemia on known substrates and cofactors of the glycolytic pathway. J Biol Chem 1964; 239:18-30.
92. Lushchak, V. I., Lushchak, L. P. Mota, A. A., Hermes-Lima, M. Oxidative stress and antioxidant defences in goldfish Carassius auratus during anoxia and reoxygenation. Am J Physiol 2001; 280: R100-R107.
93. Lutz PL. Interaction between hypometabolism and acid-base balance. Can J Zool 1989; 67:3018-3023.
94. Lutz PL. Anoxic defense mechanisms in the vertebrate brain. Ann Rev Physiol 1992; 54:601-618.
95. Lutz PL, Edwards R, McMahon P. GABA concentrations are maintained in the anoxic turtle brain. Am J Physiol 1985a; 249:R372-R374.
96. Lutz, PL , Manuel L. Maintenance of adenosine A_1 receptor function during long term anoxia in the turtle brain. Am J Physiol1999; 276:R633-R636
97. Lutz PL, McMahon P, Rosenthal M, Sick TJ. Relationships between aerobic and anaerobic energy production in turtle brain in situ. Am J Physiol 1984; 247:R740-R744.
98. Lutz PL, Nilsson G. Metabolic transitions to anoxia in the turtle brain: the role of neurotransmitters. In: Bicudo E, Glass M ed(s). The Vertebrate Gas Transport Cascade. Boca Raton: CRC Press,1993: 323-329.

99. Lutz PL, Kabler SA. Upregulation of $GABA_A$ receptor during anoxia in the turtle brain. Am J Physiol 1995; 37: R1332-R1335.
100. Lutz PL, Nilsson GE. Contrasting strategies for anoxic brain survival – glycolysis on or off. J Exp Biol 1997; 200:411-419.
101. Lutz PL, Rosenthal M, Sick TJ. Living without oxygen: turtle brain as a model of anaerobic metabolism. Mol Physiol 1985b; 8:411-425.
102. Magistretti PJ, Hof PR, Martin JL. Adenosine stimulates glycogenolysis in mouse cerebral cortex: a possible coupling mechanism between neuronal activity and energy metabolism. J Neurosci 1986; 6:2553-2562.
103. Mathers DA. The $GABA_A$ receptor: new insights from single channel recording. Synapse 1987; 1:96-101.
104. McDougal DB Jr, Holowach J, Howe MC, Jones EM, Thomas CA. The effects of anoxia upon energy sources and selected metabolic intermediates in the brains of fish, frog and turtle. J Neurochem 1968; 15:577-588.
105. McGinty D, Szymusiak R. Keeping cool: a hypothesis about the mechanisms and functions of slow-wave sleep. TINS 1990; 13:480-487.
106. McKean TA, Rocklage A, Schneider RJ: Glibenclamide binding in vertebrate cardiac membranes. J. Exp. Biol. 1993; 182: 275-281.
107. Menendez N, Solis JM, Herreras O, Sanchez Herranz A, Martin del Rio R. Role of endogenous taurine on the glutamate analogue induced neurotoxicity in the rat hippocampus. J Neurochem 1990; 55:714-717.
108. Milby KH, Mefford IN, Chey W, Adams RN. In vitro and in vivo depolarization coupled efflux of ascorbic acid in rat brain preparations. Brain Res Bull 1981; 7:237-242.
109. Milton, S. A., Lutz PL. Low extracellular dopamine levels are maintained in the anoxic turtle brain.J. Cereb. Blood Flow and Metab. 1998; 18:803-807.
109a Milton, S.L., J.W.Thompson, P.L. Lutz. Mechanisms for maintaining extracellular glutamate in the anoxic turtle striatum. Am J Physiol 2002, in press.
110. Nakada T, Hida K, Kwee IL. Brain pH and lactate acidosis: quantitative aspects of taurine effect. Neurosci Res 1992; 15:115-123.
111. Newby AC, Worku Y, Meghji P, Nakazawa, M, Skladanowski AC. Adenosine: A retaliatory metabolite or not? News Physiol Sci 1990; 5:67-70.
112. Nilsson GE. A comparative study of aldehyde dehydrogenase and alcohol dehydrogenase activities in crucian carp and three other vertebrates: apparent adaptations to ethanol production. J Comp Physiol 1988; 158:479-485.
113. Nilsson GE. Effects of anoxia on serotonin metabolism in crucian carp brain. J Exp Biol 1989; 141:419-428.
114. Nilsson GE. Long-term anoxia in crucian carp: changes in the levels of amino acid and monoamine neurotransmitters in the brain, catecholamines in chromaffin tissue, and liver glycogen. J Exp Biol 1990/ 150:295-320.
115. Nilsson GE. Turnover of serotonin in brain of an anoxia tolerant vertebrate, the crucian carp. Am J Physiol 1990b; 258:R1308-R1312.
116. Nilsson, GE. Distribution of aldehyde dehydrogenase and alcohol dehydrogenase in summer acclimatized crucian carp (*Carassius carassius* L.). J Fish Biol 1990c; 36:175-179.
117. Nilsson GE. The adenosine receptor blocker aminophylline increases anoxic ethanol excretion in the crucian carp. Am J Physiol 1991; 261:R1057-R1060.
118. Nilsson GE. Evidence for a role of GABA in metabolic depression during anoxia in crucian carp (*Carassius carassius*). J Exp Biol 1992; 164:243-259.
119. Nilsson, G. E. 2001. Surviving anoxia with the brain turned on. News Physiol. Sci. 16:217-221.
120. Nilsson GE, Alfaro AA, Lutz PL. Changes in turtle brain neurotransmitters and related substances during anoxia. Am J Physiol 1990; 259:R376-R384.
121. Nilsson GE, Hylland P, Löfman CO. Anoxia and adenosine induce increased cerebral blood flow rate in crucian carp. Am J Physiol 1994; 267:R590-R595.

122. Nilsson GE, Lutz PL. Release of inhibitory neurotransmitters in response to anoxia in turtle brain. Am J Physiol 1991; 261:R32-37.
123. Nilsson GE, Lutz PL. Adenosine release in the anoxic turtle brain as a mechanism for anoxic survival. J Exp Biol 1992; 162:345-351.
124. Nilsson GE, Winberg S. Changes in the brain levels of GABa and related amino acids in anoxic shore crab (*Carcinus meanus*). Am J Physiol 1993: R733-R737.
125. Nilsson GE, Lutz PL. Role of GABA in hypoxia tolerance, metabolic depression and hibernation - possible links to neurotransmitter evolution. Comp Biochem Physiol 1993; 105C:329-336.
126. Nilsson GE, Lutz PL, Jackson TL. Neurotransmitters and anoxic survival in the brain: a comparison between anoxia tolerant and anoxia intolerant vertebrates. Physiol Zool 1991; 64:638-652.
127. Nilsson GE, Rosén P, Johansson D. Anoxic depression of spontaneous locomotor activity in crucian carp quantified by a computerized imaging technique. J Exp Biol 1993; 180:153-162.
128. Ninomiya H, Taniguchi T, Kameyama M, Fujiwara M. Increased binding of [^{3}H] muscimol and [^{3}H] flunitrazepam in the rat brain under hypoxia. J Neurochem 1988; 51:1111-1117.
129. O'Niell RD, Fillinez M, Sundstrom L. Voltametrically monitored brain ascorbate as an index of excitatory amino acid release in the unrestrained rat. Neurosci Lett 1984; 52:227-233.
130. Okamato K, Kimura H, Sakai Y. Taurine-induced effects of the Cl^-conductance of cerebellar Purkinje cell dendrites in vitro. Brain Res 1983; 259:319-323.
131. Pannevis MC, Houlihan DF. The energetic cost of protein synthesis in isolated hepatocytes of rainbow trout (Oncorhynchus mykiss). J Comp Physiol 1993; 162B:393-400.
132. Pek M, Lutz PL. Role for adenosine in "channel arrest" in the anoxic turtle brain. J.Exp.Biol. 200, 1997, 1913-1917
133. Pek-Scott M, Lutz PL. K^+ATP channel activation provides transient protection in anoxic turtle brain. Am J Physiol 1998; 275:R2023-R2027.
134. Penney DG. Effects of prolonged diving anoxia on the turtle, *Pseudemys scripta elegans.* Comp Biochem Physiol 1974; 47A:933-941.
135. Penney D, Papademas K. Effect of starvation on fructose diphosphate, glucose-6-phosphate and phosphoglucomutase activities in organs of *Pseudemys* (*Chrysemys*) *scripta elegans.* Comp Biochem Physiol 1975; 50B:137-143.
136. Pérez-Pinzón M, Rosenthal M, Lutz PL, Sick, TJ. Anoxic survival in the isolated turtle cerebellum. J Comp Physiol 1992; 16:345-351.
137. Pérez-Pinzón MA, Chan CY, Rosenthal M, Sick TJ. Membrane and synaptic activity during anoxia in the isolated turtle cerebellum. Am J Physiol 1992; 263:R1057-R1063.
138. Pérez-Pinzón MA, Rosenthal M, Sick TJ, Lutz PL, Pablo J, Mash D. Down-regulation of sodium channels during anoxia: A putative survival strategy of turtle brain. Am J Physiol 1992; 262:R712-R715.
139. Pérez-Pinzón M, Lutz PL, Sick TJ, Rosenthal M. Adenosine, a "retaliatory" metabolite, promotes anoxia tolerance in turtle brain. J Cereb Blood Flow Metab 1993; 13:728-732.
140. Perouansky M. Pearce RA. Is anesthesia caused by potentiation of synaptic or intrinsic inhibition? Recent insights into the mechanisms of volatile anesthetics. J. Basic Clin. Physiol. Pharmacol. 2000; 11:83-107.
141. Prince DA, Stevens CF. Adenosine decreases neurotransmitter release at central synapses. Proc Natl Acad Sci USA 1992; 89:8585-8590.
142. In: Kogure K, Hossman KA, Siesjö BK eds. Molecular Mechanisms of Ischemic Brain Damage: Progress in Brain Research. Amsterdam: Elsevier Science,1985: 29-37.
143. Raffin CN, Harrison M, Sick TJ, Rosenthal M. EEG suppression and anoxic depolarization: Influences on cerebral oxygenation during ischemia. J Cereb Blood Flow Metab 1991; 11:407-415.
144. Rausch RN, Crawshaw LI, Wallace HL. Effects of hypoxia, anoxia, and endogenous ethanol on thermoregulation in goldfish, Carassius auratus. Am J Physiol 2000; 278:R545-R555.

145. Reipschläger A, Nilsson GE, Pörtner HO. A role for adenosine un metabolic depression in the marine invertebrate *Sipunculus nudus*. Am J Physiol 1997; 272:R350-R356.
146. Rice ME, Nicholson C. Interstitial ascorbate in turtle brain is modulated by release and extracellular volume change. J Neurochem 1987; 49:1096-1104.
147. Rice ME, Nicholson C. Glutamate- and aspartate induced extracellular potassium and calcium shifts and their relation to those of kainate, quisqualate and N-methyl-D-aspartate in the isolated turtle cerebellum. Neuroscience 1990; 38:295-310.
148. Rice ME, Cammack J. Anoxia-resistant turtle brain maintains ascorbic acid contents in vitro. Neurosci Lett 1991; 132:141-145.
149. Rice ME, Lee EJK, Choy Y. High levels of ascorbic acid, not glutathione, in the CNS of anoxia tolerant reptiles contrasted with levels in anoxia-intolerant species. J Neurochem 64, 1790-1799, 1995
150. Rubinsky B, Mattioli M, Arav A, Barboni B, Fletcher GL. Inhibition of Ca^{2+} and K^+ currents by "antifreeze" proteins. Am J Physiol 1992; 262:R524-R545.
151. Rudolphi KA, Schubert P, Parkinson FE, Fredholm BB. Adenosine and brain ischemia. Cerebrovasc. Brain Metab Rev 1992; 4:346-369.
152. Russo RE, Velluti JC. Inhibitory effects of excitatory amino acids on pyramidal cells of the in virtro turtle medial cortex. Exp Brain Res 1992; 92:85-93.
153. Sakurai SY, Lutz PL, Schulman A, Albin RL. Unchanged [^{3}H]MK-801 binding and increased [^{3}H]flunitrazepam binding in turtle forebrain during anoxia. Brain Res 1993; 625:181-185.
154. Sancibrian M. Opioid and prostaglandin mechanisms involved in the effects of GABAergic drugs on body temperature. Gen Pharmacol 1991; 22:259-262
155. Schenk JO, Miller E, Gaddis R, Adams RN. Homeostatic control of ascorbate concentrations in CNS extracellular fliud. Brain Res 1982; 253:353-356.
156. Schmidt H, Wegener G: Glycogen phosphorylase in fish brain (*Carassius carassius*) during hypoxia. Trans. Biochem. Soc. 1988; 16:621-622.
157. Shoubridge EA, Hochachka PW. Ethanol: novel endproduct in vertebrate anaerobic metabolism. Science Wash D.C. 1980; 209:308-309.
158. Sick TJ, Rosenthal M, LaManna JC, Lutz PL. Brain potassium ion homeostasis, anoxia, and metabolic inhibition in turtles and rats. Am J Physiol 1982; 243:R281-R288.
159. Sick TJ, Chasnoff EP, Rosenthal M. Potassium ion homeostasis and mitochondrial redox status of turtle brain during and after ischemia. Am J Physiol 1985; 248:R531-R540.
160. Sick T, Perez-Pinzon M, Lutz PL, Rosenthal M. Maintaining coupled metabolism and membrane function in anoxic brain. In: Surviving Hypoxia: Mechanisms of Control and Adaptation. P. W. Hochachka, P. L. Lutz, T. Sick, M. Rosenthal and G.van den Thilart (Eds). Boca Raton, FL: CRC Press, 1993: 351-364.
161. Siegel G, Agranoff B, Albers RW. Basic Neurochemistry: molecular, cellular, and medical aspects. 4th ed. New York: Wiley,1989.
162. Sieklucka M, Heim C, Block F, Sontag KH. Transient reduction of cerebral blood flow leads to long lasting decreases in bicuculine induced seizures. J Neural Transm Gen Sect 1992; 88:87-94.
163. Siesjö BK. Brain Energy Metabolism. New York: John Wiley and Sons,1978.
164. Siesjö BK. Acidosis and ischemic brain damage. Neurochem Pathol 1988; 9:31-88.
165. Smith RW, Houlihan DF, Nilsson GE, Brechin JG. Tissue-specific changes in protein synthesis rates in vivo during anoxia in crucian carp. Am J Physiol 1996; 271:R897-R904.
166. Stamford JA, Kruk ZL, Millar J. Regional differences in extracellular ascorbic acid levels in the rat brain determined by high speed voltammetry. Neurosci Lett 1984; 133-138.
167. Stone TW. Receptors for adenosine and adeninine nucleotides. Gen Pharmacol 1991; 22:25-31.
168. Storey KB. Tissue-specific controls on carbohydrate catabolism during anoxia in goldfish. Physiol Zool 1987; 60:601-607.
169. Storey KB. Suspended animation: the molecular basis of metabolic depression. Can J Zool 1988; 66:124-132.

170. Storey KB. Metabolic adaptations supporting anoxia tolerance in reptiles: recent advances. Comp Biochem Physiol 1996; 113B:23-35.
171. Storey KB, Storey JM. Metabolic rate depression and biochemical adaptation in anaerobiosis, hibernation and estivation. Quart Rev Biol 1990; 65:145-174.
172. Suzue T, Wu G-B, Furukawa T. High susceptibility to hypoxia of afferent synaptic transmittion in the goldfish sacculus. J Neurophysiol 1987; 58:1066-1079.
173. Swanson RA, Choi DW. Glial glycogen stores affect neuronal survival during glucose deprivation in vitro. J Cerebral Blood Flow Metab 1993; 13:162-169
174. Tanaka T, Yoshida M, Yokoo H, Mizoguchi K, Tanaka M. The role of ATP-sensitive potassium channels in striatal dopamine release: an in vivo microdialysis study. Pharmacol Biochem Behav. 1995 52, 831-835.
175. Ultsch GR, Jackson DC. Long-term submergence at 3°C of the turtle, *Chrysemys picta bellii*, in normoxic and severely hypoxic water. I. Survival, gas exchange and acid-base status. J Exp Biol 1982; 96:11-28.
176. Urenjak J, Tegtmeier F, Beile A, Khan S, Peters T. Synaptosomal respiration: a potent indicator of veratridine induced Na^+ influx. Pharmacology 1991; 43:26-35.
177. Van den Thillart G., Van Waarde A. pH changes in fish during environmental anoxia and recovery: the advantages of the ethanol pathway. In: Physiological strategies for Gas Exchange and Metabolism.A.J. Woakes, M. Grieshaber and C.R. Bridges (Eds). SEB Seminar Series, 1991: 173-190.
178. Van der Linden A, Verhoye M, Nilsson GE. Does anoxia induce cell swelling in carp brains? In vivo MRI measurements in crucian carp and common carp. J Neurophysiol 2001; 85:125-133
179. Van Waarde A, van den Thillart G, Verhagen M. Ethanol formation and pH -regulation in fish. In: Surviving Hypoxia: Mechanisms of Control and Adaptation. P. W. Hochachka, P. L. Lutz, T. Sick, M. Rosenthal and G.van den Thilart (Eds). Boca Raton, FL: CRC Press, 1993: 157-170.
180. Van Waversveld J, Addink ADF, Van den Thillart G: Simultaneous direct and indirect calorimetry on normoxic and anoxic goldfish. J Exp Biol 1989; 142:325-335.
181. Van Wylen DG, Park TS, Rubio R, Berne RM. Increases in cerebral interstitial fluid adenosine concentration during hypoxia, potassium infusion, and ischemia. J Cereb Blood Flow Metab 1986; 6:522-528.
182. Walker JM, Berger RJ. Sleep as an adaptation for energy conservation functionally related to hibernation and shallow torpor. Prog Brain Res 1980; 53: 255-278.
183. Walker RM, Johansen PH. Anaerobic metabolism in goldfish (*Carassius auratus*). Can J Zool 1977; 55:1304-1311.
184. Wasser JS, Jackson DC. Effects of anoxia and graded acidosis on the levels of circulating catecholamines in turtles. Respir Physiol 1991; 84:363-377.
185. Wasser JS, Warburton SJ, Jackson DC. Extracellular and intracellular acid-base effects of submergence anoxia and nitrogen breathing in turtles. Respir Physiol 1991; 83:239-252.
186. Willmore WH, Storey KB. Antioxidant systems and anoxia tolerance in a freshwater turtle, *Trachemys scripta elegans*. Mol Cell Biochem 1997; 170:177-185.
187. Wilson AM, Kriegstein AR. Turtle cortical neurons survive glutamate exposures that are lethal to mammalian neurones. Brain Res 1991; 540:297-301.
188. Wood SC, Gonzales R. Hypothermia in hypoxic animals: mechanisms, mediators, and functional significance. Comp Biochem Physiol B 1996 113:37-43.
189. Xia Y, Haddad GG. Major differences in CNS sulfonylurea receptor distribution between the rat (newborn, adult) and turtle. J Comp Neurol 1991; 314:278-289.
190. Xia Y, Haddad GG. Neuroanatomical distribution and binding properties of saxitoxin sites in the rat and turtle CNS. J Comp Neuro. 1993; 330:363-380.
191. Xia Y, Haddad GG. Major difference in the expression of ∂ and μ opioid receptors between turtle and rat brain. J Comp Neurol 2001; 436:202-210.
192. Xie Y, Dengler K, Zacharias E, Wilffert B, Tegtmeier F. Effects of the sodium channel blocker tetrodotoxin (TTX) on cellular ion homeostasis in rat brain subjected to complete ischemia. Brain Research 1994; 652:216-224.

193. Yao, Z., T. Mizumura, D.A. Mei and G.J.Gross. K_{ATP} channels and memory of ischemic preconditioning in dogs: synergism between adenosine and K_{ATP} channels. Am. J. Physiol. 272: H334-H342, 1997.

194. Yang J, Isenberg KE, Zorumski CF. Volatile anesthetics gate a chloride current in postnatal rat hippocampal neurons. FASEB J 1992; 6:914-918.

195. Zapol WM, Liggins GC, Schneider RC, Qvist J, Snider MT, Creasy RK, Hochachka PW. Regional blood flow during simulated diving in the conscious Weddell seal. J Appl Physiol 1979; 47:968-973.

CHAPTER 8

SPECIAL CASES OF ENHANCED TOLERANCE

1. THE MAMMALIAN NEONATE

Numerous studies have shown that for the first few days after birth, the brain of the mammalian neonate is much more resistant to hypoxia than that of the adult.[19,99,114] For example, newborn rats can survive in a N_2 atmosphere as much as 30 times longer than adults of the same species.[19]

It is of considerable interest therefore, to compare what is known of the mechanisms that account for anoxia tolerance in the turtle and crucian carp with those in the neonate to see if there are any common features.

Without doubt a substantial part of the greater hypoxia tolerance of the neonate brain can be attributed to its lesser energy requirements. For example, the mass specific cerebral oxygen and aerobic glucose consumption for the 7 day old rat is only about one-tenth that of the adult[114] and the rate of fall in brain ATP during anoxia is much less in the neonate.[19] The activity of creatine phosphokinase is correspondingly three to six fold less in neonatal brain[83] and the concentrations of mitochondrial enzymes for ATP production are also lower.[32] The lower rate of O_2 consumption gives the neonate brain the advantage, as far as hypoxia tolerance is concerned, of producing a smaller blood-tissue O_2 gradient compared to the adult.[82] This would allow the neonate to have a lower systemic PO_2 for the same tissue PO_2 as the adult. However, the mitochondria of neonate and adult brain *in situ* appear similarly sensitive to reductions in blood oxygen content in that they show a critical decline in the mitochondrial phosphorylation potential (defined as a fall in the brain PCr/P_i ratio by 50%[82]) at the same brain vascular

---

The Brain Without Oxygen. Third Edition. Peter L. Lutz, Göran E. Nilsson and Howard M. Prentice.

oxygen saturation levels. In the neonate, however, this will occur at a lower arterial PO_2 due to the higher oxygen affinity of neonate blood.[82]

During anoxia the brain will become dependent upon anaerobic glycolysis for its energy supply. Inhibition of anaerobic glycolysis by iodoacetate accelerates ATP depletion in the anoxic cerebral cortex of 2 week old rats[5] and causes a blockade of electrical respiratory activity in the brain stem of the neonate rat (which can persist during 60 min anoxia) followed by a loss of ionic homeostasis.[2] However, the glycogen stores of 2 day old neonatal rats (4 μmol glucosyl units/g[113]) are not greater than those of anoxia tolerant vertebrates (see Fig. 8.1, chapter 8), but they are rapidly consumed during anoxia.[113] In neonatal mice plasma glucose concentrations increase during anoxia[46] and hypoxia increases local cerebral blood flow in newborn dogs[19] thereby increasing the supply of glycolytic substrate.

Corresponding to an increased use of glucose there is an increase in the levels of expression of the 2 primary facilitative glucose transporter proteins in brain, GLUT1 (blood-brain barrier and glia) and GLUT3 (neuronal), during neonate maturational.[116] Interestingly, while chronic hypoxia induced an increase in GLUT1 mRNA in fetal brain, a decrease was found in hypoxic adult brain.[125]

There is much evidence that the pretreatment of neonates with glucose appears to prolong hypoxic survival and may even reduce permanent brain damage.[115] For example, in the isolated neonatal rat brainstem a high glucose concentration (30 mM) in the perfusate protects against anoxic disturbances of ion homeostasis.[117] Indeed Vanucci et al.[115] argue that glucose therapy during the labor and delivery of asphyxiated human infants might substantially reduce hypoxic-ischemic brain damage.

In the newborn the pattern of changes in extracellular K^+ of the hypoxic brain are very similar to those seen other vertebrates viz., a slow (phase 1) increase in extracellular K^+ followed by a fast (phase 2) rise which saturates at about 50-80 mM K^+(Fig. 9.1).Younger newborns require higher extracellular K^+ levels to trigger phase 2, viz. 25mM K^+ for 4-12 day olds, but the threshold declines to about 10 mM K^+ for animals more than 10 days old.[36] The rates of increase of extracellular K^+ in phase 1 are also much slower in the hypoxic newborn mammal brain compared to the adult[35,36,112] with the result that the time to reach the depolarization triggering threshold is greater the younger the animal

(Fig. 8.1). As we saw earlier, a greatly extended phase 1 is characteristic of the anoxia tolerant turtle and crucian carp brain.

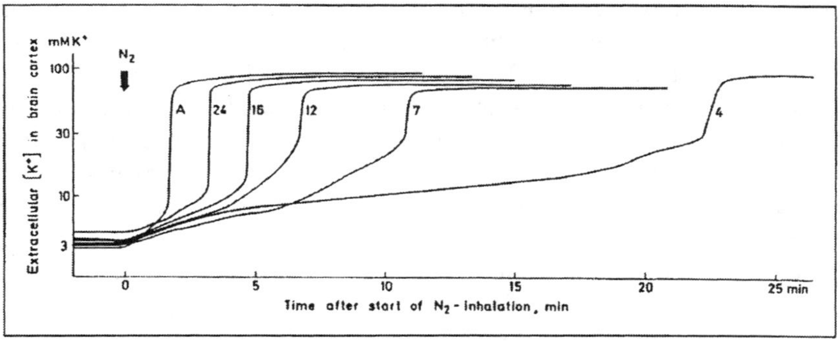

Fig. 8.1 Changes in extracellular K^+ in the brain cortex of rats at different ages following exposure to N_2. The number indicates the age (days) of the animal. A stands for adult. From Hansen.[36]

Kass & Lipton[51] reported that rats as old as 30-40 days still show an enhanced hypoxia tolerance. Studying hippocampal slices from such animals, they found that the evoked potential spike in dentate granule cells recovered 78% of its preanoxic amplitude after a 10 min anoxic episode. In the adult the recovery was only 4%. They linked the enhanced anoxic survival of the newborn to its greater ability to maintain ATP above a critical level below which irreversible synaptic failure occurs, but suggested that the actual critical level is independent of age.

However, it probable that the hypoxia tolerance of the neonate brain is not simply a matter of a lesser metabolic rate with its consequent slower fall in brain ATP levels and increase in extracellular K^+. For example, early in neonatal development there are major changes in ion channel distribution and in receptor properties, that are likely to have strong influences on hypoxia tolerance. By studying the effect of a specific blocker of mammalian K_{ATP}-channels (glibenclamide) on ion homeostasis[50] obtained results suggesting that about 50% of the increase in extracellular K^+ in the anoxic adult rat brain is caused by the opening of K_{ATP}-channels, while no signs of an opening of K_{ATP}-channels could be found in the brain of anoxic newborn rats. It appears that the neonate brain has a much lesser density of K_{ATP} channels compared to the adult[124], which may be a major factor in the much slower rate of increase of extracellular K^+ seen in the anoxic neonate brain.[50] After birth there is also a very substantial increase in the Na^+ channel density of the brain and a change in the ratio of the R_I and R_{II} Na^+ channel subtypes, which have distinct functional properties.[10,29] In the rat forebrain, sodium channel density increases five fold between birth and 28 days postpartum.[98] Differences in calcium channel modulation may also be involved. The anoxia-induced increase in intracellular Ca^{2+} is very much slower in the cerebral cortex slices of two week old rats compared to adults, and the neonate brain slice is less sensitive to glutamate evoked increases in intracellular Ca^{2+}.[5]

The brains of immature animals in general have much higher levels of taurine than the adult brains[47] and it has been suggested that this may protect the neonate brain by attenuating the harmful accumulation of intracellular Ca^{2+} that follows ischemia/anoxia.[99] High taurine may also protect the anoxic neonate brain from destructive pH changes. Using NMR and in vitro acid titration, Nakada et al.[77] showed that high levels of cystolic taurine substantially increase the acid buffering capacity of brain. However, the most ubiquitous action of taurine is as an osmoregulator[47] which may be important for survival since the brain is particularly vulnerable to osmotic disturbances during and following anoxia and ischemia.[47]

The release of adenosine in the hypoxic neonate brain may play a role in the hypoxic survival[41,88], but the immaturity of the adenosine receptors[14] and the comparatively small increases in adenosine concentrations[90] indicate that the protective effects may be minor. Indeed, xanthine and hypoxanthine (two degradation product of purines like adenosine), accumulate to such an extent in the hypoxic fetal sheep cerebral cortex[54] and in the extracellular fluid of the neonate rat striatum[90], that they "set the stage" for the generation of damaging free radicals when O_2 is restored.[54]

The characteristic pattern of anoxia induced changes in the concentrations of brain neurotransmitters/neuromodulators seen in anoxia tolerant species, viz. an increase of the concentrations of putative inhibitory amino acids (GABA, glycine, taurine) combined with a decrease in the excitatory amino acid glutamate (and its precursor glutamine)[81], is not seen in any of the major brain regions of the mammalian neonate.[68] Indeed, each region shows a different set of changes following 20 min anoxia, viz., a decrease in aspartate in the brain stem, increases in GABA and alanine and a decrease in aspartate in the telencephalon, an increase in glutamate and GABA in the diencephalon, and in the cerebellum, enhanced glycine and alanine levels.

Also unlike turtles, but similar to adult mammals, in newborn pups hypoxia causes a release of dopamine into the extracellular fluid[30] – an important cause of cellular damage in adults.[28] In early post-natal rats hypoxia caused long lasting changes in both dopamine release and the dopamine uptake mechanisms.[48] It was postulated that this long lasting effect on the striatal dopaminergic system was caused by alterations of gene expression caused by hypoxia during the critical period of the neonatal rat brain development, resulting in permanent effects on the dopamine metabolism and on the function of synaptic membranes. Moreover, even brief exposure to moderate hypoxia markedly disrupts striatal catecholamine metabolism in the immature brain.[29] It is possible then that although the immature brain can better survive hypoxia, the disruption in dopamine metabolism at crucial stages in development may cause permanent changes in synaptic function.[30,48]

However, even if the neonate brain is sensitive to glutamate toxicity[72] there is evidence that neonates do not suffer as much excitatory amino acid damage as adults after hypoxia or ischemia.[11] This is apparently not due to the absence of functional NMDA and quisqualate receptors.[49] Indeed there may even be a transient over expression of NMDA receptors in the immature rat brain.[40] The answer may lie in the circumstances that cause excitatory amino acid release. The anoxic 5 day old rat brain is protected from excitatory amino acid damage at least in part because excitatory amino acids are not released as long as the brain does not depolarize, so that delaying the depolarization delays the harmful release.[90] There may also be differences in the pattern of neurotransmitter release associated with development. Gordon et al.[31] reported that hypoxia/ischemia in the striatum of 7 day old rats produces only transient and inconsistent changes in extracellular glutamate and that the rise in glutamate is very much smaller than that seen in the adult. Indeed, Cherici and colleagues[11] found that even depolarization of the hippocampal slices from 4-9 day old newborn rats does not result in the release of glutamate or aspartate. Further protection may be provided by the neuroprotective neuromodulator taurine which is released in massively greater amounts in immature compared to adult brain following depolarization.[85]

The function of neurotransmitter receptors may also differ from those seen in mature neurons. For example, in sharp contrast to adults, benzodiazepines do not potentiate the GABA-induced inward currents in neonatal rat hippocampal cells.[87] Indeed in the early postnatal period GABA acts as an excitatory transmitter in hippocampal neurons, by promoting an increase in intracellular calcium.[11,84] The $GABA_A$ mediated depolarization could be due to immature neurons having high intracellular chloride generated by an active uptake of Cl^-.[12] Thus, in the neonate $GABA_A$ receptor activation may cause an outflow of Cl^- rather than an inflow, thereby depolarizing the neuron.

One of the protective mechanisms of survival in turtle brain, Na^+ channel down regulation[89] may also operate in the neonate brain. Dargent & Couraud[13] found that increasing the Na^+ flux in cultured brain cells from the rat fetus produced a rapid down-regulation of Na^+ channels.

To conclude, while it has been suggested that the diminished cerebral vulnerability to hypoxia of the mammalian fetus is due to functional analogies with the turtle brain[104] the most important causes of the increased hypoxia tolerance of the neonate may be just simple consequences of the comparatively undifferentiated state of the brain of the newborn, with its lower energy requirements and slower decline in ATP, lower excitability levels, delayed depolarization and lesser scope to release harmful excitatory amino acids. However, the neonate brain may also possess some "add on" features to protect against hypoxia. These could include enhanced levels of small protective molecules such as taurine as well as differences in the properties or regulation of ion-channels and receptors.

2. THE HIBERNATING MAMMAL

As we emphasized in previous chapters, all truly anoxia-tolerant animals have the ability to suppress their metabolic activities to new set points well below their usual normoxic levels. For "warm blooded" mammals the metabolic rate could be effectively reduced by allowing the body temperature to fall, and indeed mild hypothermia is being increasingly used as an effective clinical procedure to enhance ischemic and anoxic survival time (see chapter 10). Unfortunately, most mammals cannot tolerate any substantial depression of body temperature; and profound reductions in body temperature to 4-8 °C can be withstood only for minutes.[121] Forced hypothermia lowers the rate of all metabolic processes in an uncontrolled manner causing a loss of metabolic coordination, and low temperatures interfere with the normal functioning of the cell membrane and interrupt enzyme activities.[44,121] A critical factor appears to be that potassium leakage at low temperatures, which occurs chiefly through Ca^{2+} activated K^+ channels, is less dependent upon temperature than ATP-dependent K^+ accumulation.[121] However, with cardiopulmonary bypass, cardiac surgeons have been using hypothermic circulatory arrest in children since the 1950s. Body temperature is reduced to 15 °C and, as might be expected from the previous section, the neurological risk is less in neonates than in older children.[60]

Although hypothermia is a hazard to most mammals, there are striking exceptions to this rule: the mammalian hibernators which allow their body temperature to decline towards winter ambient temperatures during torpor. The body temperature of ground squirrels, for example, can range between 2-10 °C during the months of winter hibernation.[121] Not surprisingly, correspondingly great reductions in the rates of oxygen consumption and related physiological processes are seen during hibernation. In the hibernating ground squirrel the O_2 consumption is only 4-10% of normal and the heart rate drops from 500 to 20 beats/min.[44] Arterial blood pressure has been recorded as falling from 139 to 28 mm Hg.[21] During this time the breathing pattern is often slowed down or irregular. For instance, the hibernating hedgehog breaths with bursts of 40-50 breaths in 3-5 min blocks, separated by periods of apnea that average 60 min and can last as long as 150 min[59] and the hibernating garden dormouse (*Eliomys quercinus*) shows a similar pattern of apneic periods of about one hour between short respiratory bouts.[87] As might be expected, during these periods fairly severe hypoxia can be experienced. For example, in the hedgehog the average PaO_2 declined from 120 torr at the beginning of an apneic bout to 10.5 torr at the end.[109] This raises the question of whether the brain of the hibernating mammal shows any special adaptations for hypoxia tolerance.

A profound depression is seen in the brains of hibernating animals. The rates of cerebral glucose utilization in deep hibernating ground squirrels is only 1-2% of values in active animals.[22] In the hibernating squirrel cerebral blood flow is reduced by 80-90%, levels that would be considered severely ischemia in the awake animal (from 62 ml 100g min to 7 ml 100 g min).[21] Brain protein synthesis is almost halted, being reduced to 0.04% of the rate for active squirrels.[23] It has been suggested, in consequence, that hibernating mammals posses a natural tolerance to ischemia and reduced oxygen availability.[21]

Whether hibernation elicits special energy saving mechanisms in the brain is unknown, but it is likely. The EEG is severely reduces (Fig. 8.2), but the patterns of electrical activities of the mammalian brain during sleep, and during hibernation show a similar dominance of slow wave EEG patterns.[39,41]

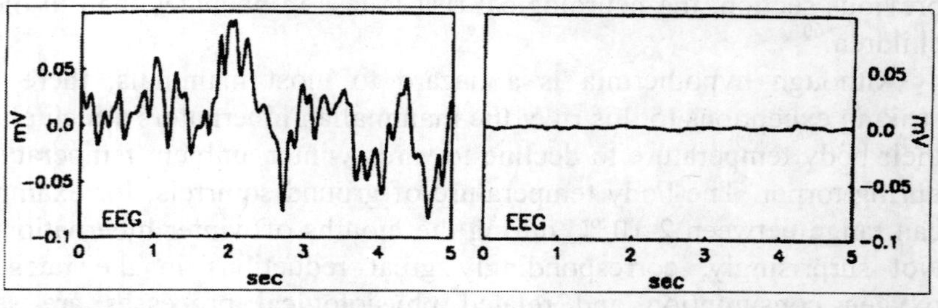

Fig. 8.2 EEG traces in awake (left panel) and hibernating (right panel) ground squirrels. From Frerichs et al.[21]

It has been suggested hibernation is entered through an extension of the slow wave stage of sleep.[42] In hibernating ground squirrels the EEG pattern associated with rapid eye movement sleep (REM) is suppressed and hibernation consists primarily of a state homologous to non-rapid-eye movement sleep interrupted periodically with short intervals of electrophysiologically defined wakefulness.[58] During normal sleep the slow wave sleep episodes are associated with brain and body cooling and a reduced cerebral metabolism.[73] During hibernation the brain blood flow may be reduced by 85% or more[21] and in addition, there is a substantial reduction in brain protein synthesis.[25] Concomitant structural alterations may also take place in the brains of hibernators. Striking changes have been seen in the mossy fiber synapses of the CA3 hippocampal neurons of hibernating ground squirrels which include a decrease in the number and size of dendritic spine infoldings.[91] Taken together these factors indicate a deep depression in brain activity and consequently, in its energy demand. Yet, although the hibernating mammal shows no behavioral responses to many external stimuli, a slight physical disturbance is often sufficient to trigger the onset of increased thermogenesis, which results in arousal after an hour or more. Apparently, during deep hibernation some regions of the brain must remain in an alert, active state.

There is increasing evidence that neurotransmitters play key roles in the down-regulation of neuronal activity that occurs during the entry to hibernation, in the maintenance of the hibernation state and in the initiation of arousal.

In the previous chapter we discussed the possible role of increased GABA levels in metabolic depression. With regards to changes in brain GABA levels, striking similarities between anoxic metabolic depression and hibernation has been pointed out by Nilsson & Lutz.[81] Thus, in ground squirrels, hamsters and hedgehogs there are considerable increases in the brain GABA levels during cold hibernation.[1,74] For example, in hibernating hamsters the brain GABA level is about twice that of nonhibernating individuals, while during arousal the GABA levels fall back to prehibernation levels.[66] But on the other hand, a decrease in striatal extracellular GABA has been reported for hibernating arctic ground squirrels.[86]

However, experimental evidence that functionally ties the changes in GABA to hibernation is lacking. Nevertheless, as mentioned in the previous chapter, injections of GABA or a GABA receptor agonist into rats cause the body temperature to decrease by several °C (see Sancibrianet et al.[97] for review). It is very possible that this inhibitory neurotransmitter has similar effects in the endogenous control of body temperature in hibernating mammals.

The monoamines have also been implicated in the hibernation process since the concentrations of many of them change during this period.[3,118] The brain monoamines are known to be involved in thermoregulation in mammals and possibly in the regulation of sleep.[118] Interestingly, during the initiation of hibernation significant increases are seen in cortical serotonin levels of the ground squirrel[34] while a fall in the levels of cortical dopamine and serotonin occur in the little brown bat during arousal from hibernation[3] indicating a role for serotonin and dopamine in coordinating these changes of state.

Interestingly like the turtle in anoxia arctic ground squirrels increases levels of the antioxidant ascorbic acid during hibernation, 3-4-fold in plasma and double in cerebrospinal fluid.[17]

However, a basic question still remains to be answered: is the neural depression seen during hibernation just a reflection of the reduced body temperature, or does the brain go down even further in its rate of metabolism? Enzymatic activity rates of hibernators appears to have Arrhenius Q_{10} values of around 2.3 which would indicate a 12 fold depression in resting metabolic rate when the body temperature was reduced from 37 ºC to 7 ºC. But the whole body metabolic depression for many hibernators is much greater, being about 20 fold for the ground squirrel *Spermophilus*, which suggests a metabolic inhibition is in effect quite beyond the simple effect of temperature.[71] The metabolic depression during hibernation must be achieved through a coordinated downregulation of key metabolic processes.

Clearly, a knowledge of the mechanisms that initiate and maintain metabolic depression in hibernating mammals may prove valuable for designing intervening strategies for brain anoxia in non-hibernators such as man.

3. THE MARINE MAMMAL

While a human will suffocate if breathing is stopped for more than 3-4 minutes, some marine mammals can breathhold dive for astonishing lengths of time. For instance, female elephant seals make dives of up to two hours duration[43] continuously for months at a time[65] and Weddell seals can remain submerged, actively swimming, for 1 hour or so.[7] This extraordinary feat is accomplished by marine mammals having special adaptations for extended apnea.

One of the most important adaptations is an increase in the amount of oxygen the animal carries at the beginning of a dive. For example, in the Weddell seal the average oxygen content of the body is 60 ml per kg body mass[57] compared to 10-15 ml per kg body mass in humans.[56] For the deep diving marine mammals the enhanced oxygen stores are the result of larger blood volumes as well as elevated hematocrits (64% Northern elephant seal, 45% man) and increased tissue myoglobin concentrations (24-81 mg per g muscle tissue for seals and whales compared to 3.0 to 7.8 for terrestrial species[105]).

For some species dive endurance is also extended through adaptations that produce a dramatic alteration in the distribution and rate of blood flow, hypoperfused tissues sharply reducing their metabolic rates.[67] Diving initiates a profound peripheral ischemia cutting off blood and oxygen supply to the muscles, liver and kidneys but maintaining blood flow for the more hypoxic vulnerable tissues, the brain and the heart.[121] Indeed, in the sea lion the cerebral blood flow even increases by as much as 120% during short voluntary dives.[16] In order to maintain blood pressure the reduction in cardiac output is accompanied by a proportionate bradycardia, which can be very substantial. For example, during a voluntary dive the gray seal heart rate can drop from 119 beats/min to as low as 4 beats/min.[111]

Marine mammals are notable in possessing well developed brains that are proportionately much larger than those of most mammals. Indeed the brain weight/body weight ratio of the dolphin is close to that of humans and their cerebral hemispheres have an even higher index of surface convolution (2.86 for human brains, 4.47 for the Bottlenose dolphin of the same size[95]). Indeed, the average cortical surface area of the dolphin brain is 3,745 cm^2 compared to 2,275 cm^2 for humans.[86] The synaptic density of the visual cortex of the dolphin is about 80% greater than that of man.[27] Their brains also have similar energy demands, the brain glucose consumption of the Weddell seal, 0.3 µmol/g. min[76], being comparable to that of the human, 0.27 µmol/g.min.[102]

One might therefore, expect from its complexity that the marine mammal brain is as equally susceptible to hypoxia as other mammals including man. However, although the experimental data is scanty there is some evidence that the marine mammalian brain has an enhanced hypoxia tolerance. The self-induced hypoxia that occurs during prolonged breathhold dives can be quite substantial. For example, in free diving Weddell seal PaO_2 may fall to 18 torr and similar values have been recorded in the Harbor seals.[128] For man short term memory is affected when the arterial oxygen falls below 55 Torr and unconsciousness occurs at about 30 torr.[102] In the rat brain energy failure, a fatal event if it lasts for more than a few minutes, occurs at a PaO_2 of about 20 torr.[102] Kerem & Elsner[53] reported that in the Harbor seal, gross signs of cerebral malfunction (seen as a severely altered EEG rhythm) did not occur until the PaO_2 reached 10 torr and the PvO_2 2.6 torr, much lower than the corresponding values for the dog ($PaO_2 = 14$ torr, $PvO_2 = 9$ torr). This level of hypoxia must be below the lowest that can be tolerated since a diving animal that became unconscious would surely drown.

Part of the enhanced brain hypoxia tolerance may be due to the microstructure of blood supply to the marine mammal brain. Compared to terrestrial mammals the dolphin cortex has more densely packed loops of capillary networks, each capillary loop encircling a small group of neurons.[26] The result of this anatomical arrangement is that the cylinder of brain tissue supplied by one capillary in the dolphin cortex is only 9-10 μm compared to 30 μm for man and 37 μm for the horse.[26] Other factors being equal the greater capillary density of the dolphin brain would produce a lesser capillary to tissue oxygen gradient compared to land mammals, and allow lower systemic O_2 tensions for the same critical cellular PO_2.

There are also some indications that the marine mammal brain has an enhanced glycolytic capacity. The glycogen content of the seal brain is some 3 times higher than that of the land mammal: Weddell seal cerebral cortex containing 8-10 μmol hexose/g, while that of the rat contains 3 μmol hexose/g.[52] Furthermore 20-25% of the brain glucose uptake of the Weddell seal is released as lactate even during air breathing[76] compared to less than 10% for the human brain[102] indicating a greater glycolytic component in aerobic energy production in seals compared to humans. In fact, there is evidence of a 40% increase in glucose consumption in the Weddell seal brain during forced dives.[76] Although brain glycolytic enzyme activities appear similar between marine and terrestrial mammals[9], seals and whales show relatively higher proportions of M-type LDH.[101] M-LDH is less sensitive to pyruvate inhibition and will operate more efficiently during anaerobic glycolysis. The high levels of M-LDH also facilitates the switch from lactate release to lactate uptake in the Weddell seal brain during recovery from a dive, when blood lactate rises above 6 mM – i.e. at lactate concentration high enough to support the normal metabolism of this organ.[76] Increases in brain pyruvate kinase activities are also seen in marine mammal species (sea lion, harbor seal, Weddell seal), and have been associated with higher maximal diving times.[92]

Another interesting feature of the marine mammal brain concerns the nature of the blood brain barrier. While in most vertebrates the blood-brain barrier is formed from endothelial cells interlocked with tight junctions,[93] dolphins and whales appear to possess a unique blood-brain interface comprising of both endothelial and glial cells.[26] These form two distinct concentric rings of tissue and it has been suggested that this (presumably) enhanced blood brain barrier in the hypoxic brain protects it from harmful systemic changes.[26]

The extraordinary abundance of glial cells surrounding neurons and vessels in the dolphin cerebral cortex[26] may also play a particularly important role in protecting the marine mammal brain during hypoxia. In the mammal almost all of the brain glycogen is located in the glial cells (astrocytes) and altering glial glycogen stores affects neuronal survival during glucose deprivation.[108] The increased glycogen content of the marine mammal brain may, therefore, reflect an enhanced capacity for their glial cells to support neuronal glycolysis under hypoxic conditions. Since glial cells are largely responsible for the uptake of extracellular K^+ lost from neurons, the increased density of these cells in the marine mammal brain may slow the hypoxia induced rise in extracellular K^+[24] and thus (similar to that seen in the brain of the newborn mammal) increase hypoxic tolerance by extending the time to reach the critical threshold for extracellular K^+ which provokes depolarization. Glial cells also play an important role in the regulation of the extracellular fluid osmolarity and in the uptake and catabolism of neurotransmitters[24] all of which are perturbed in the mammalian brain during hypoxia.

And finally, there is the curious observation that resting dolphins can have one half of their brain asleep while the other half is active, and then switch sides, allowing the dolphin to remain alert during a full sleeping bout.[64] If this ability could be expressed during a dive, then it would reduce the brains energy demands and thereby increase hypoxia tolerance.

4. HIGH ALTITUDE HYPOXIA

While hypoxia accommodating adaptations have evolved over many millions of years in marine mammals, in land mammals more immediate responses are seen during chronic hypoxia exposure (weeks-generations), naturally encountered when sea level species move to high altitude.

One response is an increased resistance to brain hypoxia in mammals exposed to hypobaric hypoxia In rats prolonged hypoxia causes an increase in brain capillary density which presumably would result in an improved oxygen delivery[15,57] and a tolerance of lower PaO_2 for the same rate of oxygen consumption. Rats held under conditions of hypobaric hypoxia (PaO_2 = 48 torr) for 3 weeks show an increased density of glucose transporters in isolated cerebral microvessels, and a corresponding increase blood-brain glucose transport.[37] One concession is that the hypoxic rat brain appears to increase its reliance on anaerobic glucose metabolism. Studies using 2-deoxyglucose showed an increase in brain regional metabolic rate for glucose, (by as much 40% in some areas (frontal motor cortex)), an increase in brain lactate and decrease in brain glycogen.[38] It is very probable that, taken together these changes: improved oxygen delivery, an upregulation of glucose transport and enhanced anaerobic glucose metabolism, act as compensatory mechanisms for the normoxic brain to accommodate prolonged hypoxia. A most interesting study was performed by Hochachka et al.[45] on brain glucose consumption in Quechula natives, indigenous to the high Andes at altitudes between 3,700-4,900 m. Positron emission tomography (PET). Measurements using [^{18}F]2-deoxy-2-fluoro-D-glucose showed that the Quechuas had a significantly lower rate of regional glucose uptake (about 10-20%) compared to a control group of lowland US males. The differences were most pronounced in the frontal cortex and angular gyre and less so in more primitive brain structures such as the cerebellum and the sensory cortex. Hochachka et al.[45] suggest that the lower rates of brain glucose uptake in the Quechuas may be a hypometabolic adaptation to chronic hypoxia, but paradoxically their brain metabolic rates had further decreased after a period of three weeks at low altitude. Whatever the reasons, the large and labile differences in human brain metabolic rates are intriguing and demand further study.

Neurotransmitter related changes also appear to occur in the hypobaric mammalian brain. Berndt et al.[4] found a 50% increase in the V max of the low affinity site of the dopamine transporter in the rat striatum after 14 h hypobaric hypoxia (11% O_2) and suggested that this increased capacity in the dopamine carrier represented an adaptation to hypoxia by allowing a faster removal of dopamine.

Birds in general have much greater tolerance of hypoxia than mammals, and can better handle higher altitude conditions.[69] For example; whereas unacclimated humans can suffer from acute altitude sickness at heights of 5,000 m[20], one bird, the Ruppell's griffin, has been reported flying at 11,300 m.[63] At that altitude the ambient PO_2 is only about 25% that at sea level. One advantage that birds have over mammals for tolerating hypoxia is that birds can withstand more pronounced hypocapnia (e.g. $PaCO_2 < 8$ torr have been seen in both the Peking duck and bar-headed goose[6]). A lower $PaCO_2$ would allow birds to have comparatively higher PaO_2 levels than mammals at the same ambient oxygen tensions.[92] One reason that birds are able to tolerate low PCO_2 is because, unlike mammals where hypocapnia produces a strong cerebral vasoconstriction with a resultant reduction in cerebral blood flow (in man 10 Torr $PaCO_2$ decreases cerebral blood flow by appox. 50%[123]), a fall in $PaCO_2$ has no significant affect on cerebral blood flow in birds.[33] For the mammal, the unfortunate result is that hypocapnic vasoconstriction can counteract or even exceed hypoxia induced cerebral vasodilation. Clearly, the lack of hypocapnic cerebral vasoconstriction gives birds an advantage over mammals in high altitude survival of the brain.

5. THE TIDAL SHARK

Sharks and rays (elasmobranchs) are generally considered to be relatively sensitive to low oxygen levels, and most species in this group are likely to avoid hypoxic habitats. However, the epaulette shark *(Hemiscyllium ocellatum)*, which inhabits shallow reef platforms on the Great Barrier Reef in Australia is often found trapped in isolated reef tidal pools during low tide, which can become severely hypoxic. In a recent study this shark was found to be hypoxia tolerant, surviving in water with a pO_2 of 8 torr (= 5% of air saturation) for at least 3.5 h[122] without any delayed phase of neuronal apoptosis.[94] Additional studies by the Renshaw lab. (unpublished) have indicated that this species can tolerate at least an hour of complete anoxia at temperatures close to 30 °C without showing a significant drop in brain ATP levels.

So far, experiments on this shark have yielded both interesting and surprising results. Thus, the epaulette shark shows no change in the cerebral blood flow when exposed to severe hypoxia (pO_2 of 8 mmHg) for 2 hours.[107] This result is unexpected, in view of the highly stimulatory effect that hypoxia has on brain blood flow in other species, including mammals, crocodiles and common carp[106,126] as well as the anoxia tolerant vertebrates discussed in previous chapters (crucian carp and freshwater turtle).

In contrast to the hyperglycemic response of the crucian carp and freshwater turtles to anoxia, blood glucose levels remain constant during hypoxia in the epaulette shark (Renshaw lab, unpublished). This is also the case in another hypoxic tolerant shark, the dogfish.[7] A recent study has suggested that elasmobranchs (including the epaulette shark) have a relatively low rate of cerebral ATP consumption compared to other vertebrates, as indicated by a very low brain Na^+/K^+ ATPase activity.[80] The brain Na^+/K^+ ATPase activity of the sharks and rays was only one third of that of teleosts (bony fishes), which may explain why elasmobranchs can afford brains that are generally much bigger than those of teleosts.[78,80]

However, there are also similarities between the epaulette shark and other hypoxia tolerant vertebrates. Thus, the epaulette shark brain shows signs of metabolic depression in response to hypoxia. In particular, repeated hypoxic insults result in a reduction of oxidative metabolism as measured by cytochrome oxidase activity.[75]

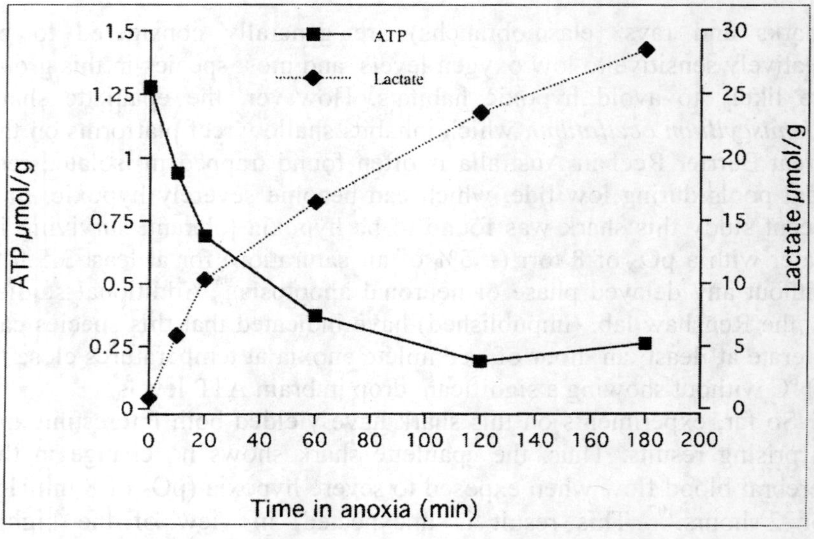

Fig. 8.3 Collapse in brain ATP levels and increase in brain lactate in frog brain during survivable anoxic exposure. Data from Wegener et al.[120]

6. THE ANOXIC FROG

And finally, the frog brain commands particular attention. The common frogs (*Rana temporaria* and *Rana pipiens*) represent an intermediate in anoxia tolerance being able to survive brain anoxia for hours at 25 °C.[70,120] Like in the true anoxia survivors we have discussed in the previous chapter, brain activity is rapidly suppressed in the anoxic frog.

In the optic tectum, spikes corresponding to ON-stimuli almost immediately decrease following anoxia and are completely lost within 10 min, i.e., the animal is functionally blind within a few min of anoxia.[120] The spinal cord is more resistant since signal transmission is preserved for up to 1 h of anoxia.[120] During this time animal activity is severely reduced and reaches a "frozen state" at 60 min and a "curare like" paralyzed state at 120 minsanoxia.[120] At the same time glycolytic flux is increased as shown by an increase in tissue lactate.

But unlike the true anoxia survivors the frog does not defend brain energy stores during anoxia (Fig. 8.3).[55,70,120] However, although brain ATP starts to decline immediately on encountering anoxia, in contrast to typical anoxia vulnerable species, frog brain ATP depletion takes hours instead of minutes.[55, 70,120] Also, while in the mammalian brain the loss of ATP results in a rapid (min) depolarization with consequent degenerative changes, the frog maintains extracellullar K^+ levels until ATP decreases to less than 35% normoxic levels, after which K^+ slowly starts to drift up.[55] After about one hour later, when extracellular K^+ reached an inflection value of about 5 mmol l^{-1}PBNI, like the mammal, there was a rapid increase in extracellular K^+ to a plateau indicative of anoxic depolarization.[55] The time to anoxic depolarization corresponds to the time for the massive loss of neurotransmitters (GABA, glutamate) in the anoxic frog brain.[70] Part of the reduction in ATP demand is due to an approximately 50% decrease in K^+ efflux in the anoxic frog brain, which would also contribute to the retarded rate of increase in extracellular K^+ during the initial phase of anoxic depolarization.[55] However, unlike the anoxia tolerant turtle brain, adenosine, which is released after about 1 hour of anoxia[66], did not appear to be involved in the down regulation of K^+ leakage in the frog.[55]

The anoxic frog brain appears to experience the same esquence of degenerative changes as the mammal but on a greatly extended time scale. The enhanced anoxic tolerance of the frog brain is more a matter of slow death rather than enhanced protective mechanisms, making it an excellent candidate model to investigate the causal sequence of degenerative events following anoxic insult in the anoxic intolerant mammalian brain.

7. REFERENCES

1. Al-Badry KS, Taha HM. Hibernation hypothermia and metabolism in hedgehogs - changes in free amino acids and related compounds. Comp Biochem Physiol 1982; 72A:541-547.
2. Ballanyi K, Kuwana S-I, Völker A. Developmental changes in the hypoxia tolerance of the in vitro respiratory networks of rats. Neurosci Lett 1992; 148:141-144.
3. Bauman WA, Hashim A, Sershen H. Cerebral cortical concentrations of bioamines and their metabolites during arousal and after feeding in the little brown bat (*Myotis lucifugis*). Brain Res 1989; 500:156-160.
4. Berndt C, Henke W, Gross J. Hypoxia induces different responses of striatal high- and low-affinity dopamine uptake sites. Mol and Chem Neuropath 1993; 18:179-187.
5. Bickler PE, Gallego SM, Hansen BM. Developmental changes in intracellular calcium regulation in rat cerebral cortex during hypoxia. J Cereb Blood Flow and Metab 1993; 13:811-819.
6. Black CP, Tenney SM. Pulmonary hemodynamic responses to acute and chronic hypoxia in two waterfowl species. Comp Biochem Physiol A 1980; 67:291-293.
7. Butler, P.J., Taylor, E.W. and Davison, W. The effect of long term, moderate hypoxia on acid base balance, plasma catecholamines and possible anaerobic end products in the unrestrained dogfish *Scyliorhinus canicula.* J. Comp. Physiol. *B.* 1979;132:297-303.
8. Castellini MA, Somero GN, Kooyman GL. Glycolytic enzyme activities in tissues of marine and terrestrial mammals. Physiol Zool 1981; 54:242-252.
9. Castellini MA, Kooyman GL, Ponganis PJ: Metabolic rates of freely diving Weddel seals: Correlations with oxygen stores, swim velocity and diving duration. J Exp Biol 1992; 165:181-194.
10. Cattertall WA. Cellular and molecular biology of voltage-gated sodium channels. Physiol Revs 1992; 72:15-48.
11. Cherici G, Alesiani M, Pelligrini-Gaimpietro DE, Moroni F. Ischemia does not induce the release of excitotoxic amino acids from the hippocampus of newborn rats. Dev Brain Res 1991; 60:235-240.
12. Cherubini E, Gaiarsa JL, Ben-Ari Y. GABA: an excitatory transmitter in early postnatal life. Trends Neurosci 1991; 14:515-519.
13. Dargent B, Courand F. Down-regulation of voltage-dependent sodium channels initiated by sodium influx in developing neurones. Proc Natl Acad Sci USA 1989; 87:5907-5911.
14. Daval J, Werck M. Autoradiography changes in brain adenosine A_1 receptors and their coupling to G proteins following seizures in the developing rat. Dev Brain Res 1991; 59:237-247.
15. Diemer K, Henn R. Kapillarvermehrung in der Hirnrinde der Ratte unter chronischem Sauerstoffmangel. Die Nature 1965; 52:135-136.
16. Dormer KJ, Denn MJ, Stone HL. Cerebral blood flow in the sea lion (*Zalophus californianus*) during voluntary dives. Comp Biochem Physiol 1977; 58A:11-18.
17. Drew KL; Osborne PG; Frerichs KU; Hu Y. Koren RE Hallenbeck JM; Rice ME Ascorbate and glutathione regulation in hibernating ground squirrels. Brain Res 1999; 851:1-8.
18. Duffy TE, Kohle SJ, Vannucci RC. Carbohydrate and energy metabolism in the perinatal rat brain: relation to survival to anoxia. J Neurochem 1975; 24:271-276.
19. Duffy TE, Cavazzuti M, Cruz NF, Sokoloff L Local cerebral glucose metabolism in newborn dogs: Effects of hypoxia and halothane anesthesia. Ann Neurol 1981; 11:233-246.
20. Folk GE. Introduction to Environmental Physiology. Environmental Extremes and Mammalian Survival. Philadelphia, PA:Lea & Febiger 1966.
21. Frerichs KU, Kennedy C, Sokoloff L, Hallenbeck JM. Local cerebral blood flow during hibernation, a model of natural tolerance to "cerebral ischemia". Cereb Blood Flow and Metab 1994; 14:193-205.
22. Frerichs KU, Dienel GA, Cruz NF, Sokoloff L, Hallenbeck JM. Rates of glucose utilization in brain of active and hibernating ground squirrels. Am J Physiol 1995; 268:R445-R453.

23. Frerichs KU; Smith CB; Brenner M; DeGracia DJKrause GS; Marrone L; Dever TE; Hallenbeck JM Suppression of protein synthesis in brain during hibernation involves inhibition of protein initiation and elongation. Proc Natl Acad Sci U S A 1998 24;95:14511-6.
24. Ghazi H, Osborne NN: Receptors on glia: their role in neuronal function. In: Winlow W ed(s). Neuronal communications. Manchester: Manchester University Press, 1990: 150-164.
25. Giacometti S, Scherini E, Bernocchi G. Seasonal changes in the nucleoli of Purkinje cells of the hedgehog cerebellum. Brain Res 1989; 488:365-368.
26. Glezer II, Jacobs MS, Morgane PJ. Ultrastructure of the blood brain barrier in the dolphin (*Stenella coeruleoalba*). Brain Res 1987; 414:205-218.
27. Glezer II, Morgane PJ. Ultrastructure of synapses and golgi analysis of neurones of the lateral gyrus (visual cortex) of the dolphin and pilot whale. Brain Res Bull 1990; 24:401-427.
28. Globus MY, Busto R, Dietrich WD, Martinez E, Valdes I, Ginsberg MD. Effect of ischemia on the in vivo release of striatal dopamine, glutamate and g-aminobutyric acid studied by intracerebral microdialysis. J Neurochem 1988; 1455-1464.
29. Gordon D, Merrick D, Auld V, Dunn R, Goldin AL, Davidson N, Catterall WA.Tissue-specific expression of the R_I and R_{II} sodium channel subtypes. Proc Nat Acad Sci 1987; 84:8682-8686.
30. Gordon K, Statman D, Johnston MV, Robinson TE, Becker JB, Silverstein FS. Transient hypoxia alters striatal catecholamine metabolism in immature brain: an in vivo microdialysis study. J Neurochem 1990; 54:605-611.
31. Gordon KE, Simpson J, Statman D, Effects of perinatal stroke on striatal amino acid efflux in rats studied with in vivo microdialysis. Stroke 1991; 22:928-932.
32. Gregson NA, Williams PL. A comparative study of brain and liver mitochondria from new-born and adult rats. J Neurochem 1969; 16:617-626.
33. Grubb B, Mills CD, Colacino JM. Effect of arterial carbon dioxide on cerebral blood flow in ducks. Am J Physiol 1977; 232:H595-H601.
34. Haak LL, Mignot E, Kilduff TS, DementWC, Heller HC. Regional changes in central monoamine and metabolite levels during the hibernation cycle in the golden-manteled ground squirrel. Brain Res 1991; 563:215-220.
35. Haddad GG, Donnelly DF. O_2 deprivation induces a major depolarization in brainstem neurones in the adult but not in the neonatal rat. J Physiol Lond 1990; 429:411-428.
36. Hansen AJ. Extracellular potassium concentration in juvenile and adult rat brain cortex during anoxia. Acta Physiol Scand 1977; 99:412-420.
37. Harik SI, Behmand RA, LaManna JC. Hypoxia increases glucose transport at the blood-brasin barrier. J Appl Physiol 1994; 77:896-901.
38. Harik SI, Lust WD, Jones SC, Lauro KL, Pundik S, LaManna JC. Brain glucose metabolism in hypobaric hypoxia. J Appl Physiol 1995; 79:136-140.
39. Harris DV, Walker JM, Berger RJ. A continuum of slow-wave sleep and shallow torpor in the pocket mouse *Perognathus longimembris*. Physiol Zool 1984; 57:428-434.
40. Hattori H, Wasterlain CG. Excitatory amino acid in the developing brain: ontogeny, plasticity and excitotoxicity. Pediatr Neurol 6:219-228.
41. Heller HC, Walker JM, Florant GL. Sleep and hibernation: electrophysiological and thermoregulatory homologies. In: Wang LCH, Hudson JW ed(s). Strategies in cold: natural torpidity and thermogenesis. London: Academic Press,1978: 225-265.
42. Heller HC. Sleep and hypometabolism. Can J Zool 1988; 66:61-69.
43. Hindell MA, Slip DJ, Burton HR. Physiological implications of continious, prolonged and deep dives of the southern elephant seal *(Mirounga leonina)*. Can J Zool 1992; 70:370-379.
44. Hochachka PW, Guppy M. Metabolic Arrest and the Control of Biological Time. Cambridge, MA: Harvard University Press, 1987.
45. Hochachka PW, Clark CM, Matheson GO, Brown WD, Stone CK, Nickles RJ, Holden JE. The brain at high altitude: Hypometabolism as a defense against chronic hypoxia? J Cereb Blood Flow and Metab 1994; 14:671-679.

46. Holowach-Thurston J, Hauhart RE, Jones EM. Decrease in brain glucose in anoxia in spite of elevated plasma glucose levels. Pediatr Res 1973; 7:691-695.
47. Huxtable RH. Physiological actions of taurine. Physiol Rev 1992; 72:101-163.
48. Ihle W, Gross J, Moller R. Effect of chronic postnatal hypoxia on dopamine uptake by synaptosomes from striatum of adult rats. Biomed Biochem Acta 1985; 44:433-437.
49. Insel TR, Miller LP, Gelhard RE. The ontogeny of excitatory amino acid receptors in the rat forebrain. I. NMDA and quisqualate receptors. Neurosci 1990; 35:31-43.
50. Jiang C, Xia Y, Haddad GG. Role of ATP sensitive K^+ channels during anoxia: Major differences between rat (newborn and adult) and turtle neurones. J Physiol 1992; 448:599-612.
51. Kass IS, Lipton P. Protection of hippocampal slices from young rats against anoxic transmittion damage is due to better maintenance of ATP. J Physiol 1989; 413:1-11.
52. Kerem O, Hammond DD, Elsner R. Tissue glycogen levels in the Weddell seal, *Leptonychotes weddelli*, a possible adaptation to asphyxial hypoxia. 1973a; 45A:731-736.
53. Kerem D, Elsner R. Cerebral tolerance to asphyxial hypoxia in the harbour seal. Respir Physiol 1973b; 188-200.
54. Kjellmer I, Andine P, Hagberg H, Thiringer K. Extracellular increases of hypoxanthine and xanthine in the cortex and basal ganglia of fetal lambs during hypoxia-ischemia. Brain Res 1989; 478:241-247.
55. Knickerbocker, D L, Lutz PL. Slow ATP loss and the defense of ion homeostasis in the anoxic frog brain. J. Exp Biol 2001;204:3547-3551
56. Kooyman GL, Castellini MA, Davis RW. Physiology of diving in marine mammals. Ann Rev Physiol 1981; 43:343-356.
57. Kooyman GL. Free diving in vertebrates. In: Dejours P ed. Comparative physiology of environmental adaptations. Basel:Karger,1987: 27-37.
58. Krilowicz BL, Glotzbach SF, Heller HC. Neuronal activity during sleep and complete bouts of hibernation. Am J Physiol 1988; 255:R1008-R1019.
59. Kristofferson R, Soivio A. Hibernation in the hedgehog (*Erinaceus europaeus* L.). Changes of respiratory pattern, heart rate and body temperature in responce to gradually decreasing or increasing ambient temperature. Ann Acad Sci Fenn 1964; 82:3-13.
60. Kurth CD, Steven JM, Nicolson MD. Cerebral oxygenation during pediatric cardiac surgery using deep hypothermic circulatory arrest. Anesthesiology 1995; 82:74-82.
61. LaManna JC, Vendel LM, Farrell R. Brain adaptation to chronic hypobaric hypoxia in rats. J Appl Physiol 1992;72:2238-2243.
62. Laudignon N, Farri E, Beharry K,Aranda JV. Influence of adenosine on cerebral blood flow during hypoxia in the newborn piglet. J Appl Physiol 1990; 68:1534-1541.
63. Laybourne RC. Collision between a vulture and an aircraft at an altitude if 37,000 feet. Wilson Bull 1974; 86:461-462.
64. Leatherwood S, Reeves RR. The Bottlenose Dolphin. Academic Press 1990.
65. Le Beouf BJ, Costa DP, Huntley AC. Continuous deep diving by nothern elephant seals. Can J Zool 1989; 66:446-458.
66. Lust WD, Wheaton AB, Feussner G, Passonneau J. Metabolism in the hamster brain during hibernation and arousal. Brain Res 1989; 489:12-20.
67. Lutz PL, Hochachka PW. Hypoxia defense mechanisms: A comparison between diving reptiles and mammals. In: Hochachka PW, Lutz PL, Sick T, Rosenthal M. ed(s). Surviving hypoxia: Mechanisms of control and adaptation. CRC Press. 1993: 459-469.
68. Lutz PL, Ortiz M, Leone-Kabler S. Regional changes in amino acid levels of the neonate rat brain during anoxia and recovery. Neurochem Res 1994; 19:1238-1287.
69. Lutz PL, Storey K. Strategies for dealing with variations in oxygen tensions by vertebrates and invertebrates. In: Handbook of Comparative Physiology: Oxford University Press. W. Danzler (ed)Am. Physiol. Soc. Oxford University Press.1997. .pp1479-1522
70. Lutz, PL and Reiners R.. Survival of energy failure in the anoxic frog brain: delayed release of glutamate. J. Exp.Biol. 1997 ,200, 2913-2917.

71. Malan A. pH and metabolic depression in mammalian hibernation: the example of brown adipose tissue. In: Hochachka PW, Lutz PL, Sick T,Rosenthal M, van den Thillart G. (eds). Surviving Hypoxia: Mechanisms of control and adaptation. Boca Raton: CRC Press,1993: 201-215.
72. McDonald JW, Silverstein FS, Johnston MV. Neurotoxicity of NDMA is markedly enhanced in developing rat central nervous system. Brain Res 1988; 459:200-203.
73. McGinty D, Symusiak R. Keeping cool: a hypothesis about the mechanisms and function of slow-wave sleep. Trends Neurosci 1990; 13:480-487.
74. Mihailovic LT, Krzalic L, Cupic D. Changes of glutamine, glutamic acid and GABA in cortical and subcortical brain structures of hibernating and fully aroused ground squirrels (*Citellus citellus*). Experientia 1965; 21:709-710.
75. Mulvey JM, Renshaw GM. Neuronal oxidative hypometabolism in the brainstem of the epaulette shark (Hemiscyllium ocellatum) in response to hypoxic pre-conditioning. Neurosci Lett 2000; 290:1-4.
76. Murphy B, Zapol WM, Hochachka PW. Metabolic activities of heart, lung and brain during diving and recovery in the Weddell seal. J Appl Physiol 1980; 48:596-605.
77. Nakada T, Hida K, Kwee IL. Brain pH and lactate acidosis: quantitative analysis of the taurine effect. Neurosci Res 1992; 15:115-123.
78. Nilsson, G. E. The cost of a brain. Natural History 1999, 108 (10): 66-73
79. Nilsson GE, Lutz PL, Jackson TL. Neurotransmitters and anoxic survival in the brain: a comparison between anoxia tolerant and anoxia intolerant vertebrates. Physiol Zool 1991; 64:638-652.
80. Nilsson, G.E., Routley, M.H, Renshaw, G.M.C Low mass specific brain Na+K+ ATPase activity in elasmobranch and teleost fishes: its implications for the large brain size of elasmobranchs. Proc Roy Soc B 2000; 267:1335-1339.
81. Nilsson GE, Lutz PL. Role of GABA in hypoxia tolerance, metabolic depression and hibernation - possible links to neurotransmitter evolution. Comp Biochem Physiol 1993; 105C:329-336.
82. Nioka S, Chance B, Smith DS, Mayevsky A, Reilly MP, Alter C, Asakura T. Cerebral energy metabolism and oxygen state during hypoxia in neonate and adult dogs. Ped Res 1990; 28:54-62.
83. Norwood WI, Ingwall JS, Norwood C. Developmental changes in creatine kinase metabolism in rat brain. Am J Physiol 1983; 224:C205-C210.
84. Obrietan K, van den Pol AN. GABA neurotransmission in the hypothalamus: reversal from Ca^{2+} elevating to depression. J Neorosci 1995; 15:5065-5077.
85. Oja SS, Saransaari P. Taurine as osmoregulator and neuromodulator in the brain. Metab Brain Disease 1996;11:2153-163.
86. Osborne PG; Hu Y; Covey DN; Barnes BN; Katz Z Drew KL Determination of striatal extracellular gamma-aminobutyric acid in non-hibernating and hibernating arctic ground squirrels using quantitative microdialysis. Brain Res 1999; 839:1-6.
87. Pajunen I. A comparison of the heart rate at different ambient temperatures during long-term hibernation in the garden dormouse *Eliomys quercinus L.* Cryobiology 1992; 29:414-421.
88. Park TS, Wylen DGV, Rubio R, et al. Increased brain interstitial fluid adenosine concentration during hypoxia in newborn piglet. J Cereb Blood Flow Metab 1987; 7:178-183.
89. Pérez-Pinzón M, Rosenthal M, Sick T, Lutz PL. Down-regulation of sodium channels during anoxia: A putative survival strategy of turtle brain. Am J Physiol 1992b; 262:R712-R715.
90. Pérez-Pinzón MA, Nilsson G, Lutz PL. Relationships between ion gradients and neurotransmitter release in the newborn rat striatum during anoxia. Brain Res 1993; 602:228-233.
91. Popov VI, Bocharova LS. Hibernation -induced structural changes in synaptic contacts between mossy fibres and hippocampal pyramidal neurons. Neurosci 1992; 48:52-62.

92. Powell, F. Acclimatization to high altitude. In: Sutton JR, Coates G, Rimmers JE ed(s). Hypoxia: The Adaptations Toronto:Decker, 1990: 41-44.
93. Rapoport SI. Blood-brain barrier in physiology and medicine. New York: Raven,1976.
94. Renshaw GM, Dyson SE.Increased nitric oxide synthase in the vasculature of the epaulette shark brain following hypoxia. Neuroreport. 1999;10:1707-1712.
95. Ridgeway SH. Dolphin brain size. In: Bryden MM, Harrison R ed(s). Research on dolphins. Oxford:Clarendon Press, 1986: 59-70.
96. Rovira C, Ben-Ari Y. Benzodiazapines do not potentiate GABA responses in neonatal hippocampal neurones. Neurosci Letters 1991; 130:157-161.
97. Sancibrian M, Serrano JS, Minano FJ. Opioid and prostaglandin mechanisms involved in the effects of GABAergic drugs on body temperature. Gen Pharmacol 1991; 22:259-262.
98. Scheinmann RI, Auld VJ, Goldin AL, et al. Developmental regulation of sodium channel expression in the rat forebrain. J Biol Chem 1989; 264:10660-10666.
99. Schurr A, Rigor B. The mechanism of neuronal resistance and adaptation to hypoxia. FEBS lett 1987; 224:4-8.
100. Schurr A, Tseng MT, West CA, Rigor BM. Taurine improves the recovery of neuronal function following cerebral hypoxia: an in vitro study. Life Sci 1987; 40:2059-2066.
101. Shoubridge EA, Carscadden JE, Legett WC. Lactate dehydrogenase isozyme patterns in cetaceans: evidence for a biochemical adaptation to diving. Comp Biochem Physiol 1976; 53B:357-359.
102. Siesjö BK: Brain Energy Metabolism. New York:John Wiley and Sons,1978.
103. Simon LM, Robin ED, Elsner R. A biochemical basis for differences in maximal diving time in aquatic animals. Comp Biochem Physiol 1974; 209-215.
104. Singer D Neonatal tolerance to hypoxia: a comparative-physiological approach. Comp Biochem Physiol A Mol Integr Physiol 1999 123:221-34.
105. Snyder GK. Respiratory adaptations in diving mammals. Resp Physiol 1983; 54:269-294
106. Söderström, V. Nilsson, G E., Renshaw, GM.C, and Franklin, CE. . Hypoxia stimulates cerebral blood flow in the estaurine crocodile (*Crocodylus porosus*). Neuroscience Letters 1999; 267: 1-4.
107. Söderström, V, Renshaw, GM.C. and Nilsson, G E. . Brain blood flow and blood pressure during hypoxia in the epaulette shark (*Hemiscyllium ocellatum*), a hypoxia tolerant elasmobranch. J. Exp.Biol. 1999; 202: 829-835.
108. Swanson RA, Choi DW. Glial glycogen stores affect neuronal survival during glucose deprivation in vitro. J Cereb Blood Flow Metab 1993; 13:162-169.
109. Tahiti H, Soivio A. Blood gas concentrations, acid-base balance and blood pressure in hedgehogs in the active state and in hibernation with periodic respiration. Ann Zool Fenn 1975; 12:188-197.
110. Tempel GE, Musacchia XJ. The hypothermic hamster brain: its water content and perfusion. Cryobiology 1981; 18:585-591.
111. Thompson D, Fedak MA. Cardiac responces of Grey seals during diving ar sea. J Exp Biol 1993; 174:139-164.
112. Trippenbach T, Richter DW, Acker H. Hypoxia and ion activities within the brain stem of newborn rabbits. J App Physiol 1990; 68:2494-2503.
113. Vannucci RC, Duffy TE. Carbohydrate metabolism in fetal and neonatal rat brain during anoxia and recovery. Am J Physiol 1976; 230:1209-1275.
114. Vannucci RC. Experimental biology of cerebral hypoxia-ischemia: relation to perinatal brain damage. Pediat Res 1990; 27:317-326.
115. Vannucci RC, Mujsce DJ. Effect of glucose on perinatal hypoxic-ischemic brain damage. Biol Neonate 1992; 62:215-224.
116. Vannucci RC, Vannucci SJ Glucose metabolism in the developing brain. Semin Perinatol 2000 Apr;24(2):107-15.
117. Völker A, Ballanyi K, Richter DW. Glucose dependence of ionic homeostasis in the isolated respiratory network of neonatal rats. Proc Int Physiol Congress Glasgow 1993; abstract 141:27.

118. Wang LCH. Time patterns and metabolic rates of natural torpor in the Richardson's ground squirrel. Can J Zool 1979; 57:149-155.
119. Wang LCH. Mammalian hibernation: An escape from the cold. In: Advances in Comparative and Environmental Physiology. Berlin: Springer-Verlag,1988: 1-43.
120. Wegener G, Krause U. Environmental and exercise anaerobiosis in frogs.In: Hochachka PW, Lutz PL, Sick T, et al. ed(s). Surviving Hypoxia: Mechanisms of control and adaptation. Boca Raton: CRC Press,1993: 217-236.
121. Willis JS. Cold tolerance in mammalian cells. Symp Soc Exp Biol 1987; 41:285-309.
122. Wise, G, Mulvey, J M Renshaw, G.M C. Hypoxia tolerance in the epaulette shark *(Hemiscyllium ocellatum)*. J Exp Zool 1998; 281:1-5.
123. Wollman H, Smith TC, Stephen GW, Colton ET 3rd, Gleaton HE, Alexander SC. Effect of extremes of respiratory and metabolic alkalosis on cerebral blood flow in man. J Appl Physiol 1968; 24:60-65.
124. Xia Y, Haddad GG. Major differences in CNS sulphonylurea receptor distribution between the rat (newborn, adult) and turtle. J Comp Neurol 1991; 314:278-289.
125. Xia Y, Warshaw JB, Haddad GG. Chronic hypoxia causes opposite effects on glucose-transporter-1 messenger RNA in mature versus immature rat-brain. Brain Res. 1995:675, 224-230
126. Yoshikawa, H., Ishida, Y., Kawata, K., Kawai, F. and Kanamori. M. Electroencephalograms and Cerebral Blood Flow in Carp, Cyprinus carpio, Subjected to Acute Hypoxia. J Fish Biol 1995; 46:114-122.
127. Zapol WM, Liggins GC, Schneider RC, Qvist J, Snider MT, Creasy RK, Hochachka PW. et al. Regional blood flow during simulated diving in the conscious Weddell seal. J Appl Physiol 1979; 47:968-973.
128. Zapol WM, Hill RD, Qvist J, Falke K, Schneider RC, Liggins GC, Hochachka PW. Arterial gas tensions and hemoglobin concentrations of the freely diving Weddell seal. Undersea Biomed Res 1989; 16:363-373.

CHAPTER 9

ACCLIMATION TO HYPOXIA IN MAMMALS: PRECONDITIONING

1. ISCHEMIC PRECONDITIONING

It has long been known that pre-exposing animals to low oxygen conditions may enhance their subsequent ability to tolerate hypoxia or even anoxia. Thus, in 1957, Prosser and co-workers showed that hypoxia exposure in the goldfish increased its ability to maintain oxygen uptake at low oxygen levels, and it also supressed its resting metabolic rate.[50] Recently, it was found that the epaulette shark, a species that we presented in the previous chapter, also displays similar respiratory adjustments when pre-exposed to hypoxia.[52] In comparative physiology, such a response is generally referred to as acclimation. During recent years, similar phenomena have been observed in mammals, and have in the context of mammalian ischemia research been termed hypoxic preconditioning, or ischemic preconditioning.

The observation that an initial brief experience of ischemia/anoxia could produce a temporary period of enhanced ischemia tolerance has provoked much research interest. Preconditioning takes place in a variety of tissues including brain, heart, liver, small intestine, skeletal muscle and lung.[13] Preconditioning has been most extensively examined in heart but without doubt determining the mechanisms that underlie myocardial preconditioning is of value in understanding how tolerance is mediated during ischemia in other tissues including brain. Ischemic preconditioning in the heart occurs in two phases, an early period of protection in the first 2-3 hours after the ischemic insult that lasts for hours and a second or late phase that starts at 12 to 24 hours after the ischemia and lasts for 3-4 days.[10,36,7]

The Brain Without Oxygen. Third Edition. Peter L. Lutz, Göran E. Nilsson and Howard M. Prentice.

2. EARLY AND LATE MYOCARDIAL PRECONDITIONING

The early phase of preconditioning (or classical preconditioning) protects against infarction and has been found to be more potent than the second window of protection.[64] Late preconditioning protects the heart against both infarction and a type of contractile dysfunction referred to as stunning.[6] In contrast early preconditioning only protects against infarction but not stunning. Another difference is that of the duration of protection which in the late phase can be extended by repetitive ischemia but in the early phase remains fixed.[11] Because late preconditioning protects against both infarction and stunning it has been argued by some investigators that that it may have greater clinical relevance than classical preconditioning.[7] Late preconditioning can be induced by a range of stimuli including heat stress[12], rapid ventricular pacing[30] and exercise[63] as well as pharmacological agents including interleukin-1 and other growth factors, reactive oxygen species and drugs such as adenosine receptor agonists and NO-releasing agents.[58,7]

3. ADENOSINE RECEPTORS

Early myocardial preconditioning can be mimicked by adenosine receptor agonists in a variety of species, and in the rabbit specifically both A1 and A3 adenosine receptor agonists may be involved.[61,2] Other preconditioning affecting agonists include bradykinin and opioids and the release of reactive oxygen species (ROS) during the initial ischemia/reperfusion may also play such a role.[44,42,3] Distinct forms of late preconditioning can be categorised in terms of the initial stimulus and the nature of the subsequent cardioprotection. For example adenosine receptor agonists induce protection against infarction[60] whereas ROS generating agents protect against stunning and infarction.[51] In addition to its role in early preconditioning adenosine is also involved in late preconditioning. Released during preconditioning ischemia, it triggers delayed protection against myocardial infarction by the activation of A1 or A3 receptors.[60] Pharmacological stimulation of either A1 or A3 receptors can elicit late preconditioning against infarction but it has not as yet been determined whether only one or both of these receptor subtypes contributes to triggering ischemia-induced late preconditioning.[7]

4. K_{ATP} CHANNELS

K_{ATP} channels may play an important role in early preconditioning. Activation of K_{ATP} channels by administration of bradykinin or cromakalim is cardioprotective.[65,25] Inhibitors of K_{ATP} channels including 5-hydroxydecanoate (5-HD) and glibenclamide will cause blockade of the protective effects of both ischemic and pharmacological preconditioning.[29,2] However, K_{ATP} channels may be critical only during the prolonged ischemic period as protection in dogs is blocked by glibenclamide administered after the preconditioning but before the 60 minute ischemia.[24] While it was originally believed that myocardial preconditioning was mediated by sarcolemmal channels this was contradicted by the observation that the K_{ATP} channel opener bimakalim reduced infarct size at a dose that did not affect action potential duration.[65] In addition the cardioprotective properties of the K_{ATP} channel opener cromakalim were independent of a shortened action potential duration. These observations point to the conclusion that the protective effects of these K_{ATP} channel openers were not mediated through sarcolemmal K_{ATP} channels. The K_{ATP} channel blocker 5-HD which inhibits preconditioning failed to inactivate chromakalim-induced K_{ATP} currents, again contradicting a role for sarcolemmal K_{ATP} channels in the early phase of preconditioning.[41] By contrast, recent evidence indicates that 5-HD selectively blocks mitochondrial K_{ATP} channels.[18] Downey and colleagues have demonstrated that the selective mitochondrial K_{ATP} channel opener diazoxide reduces infarct size in anesthetised rabbits and that this protection was reversed with 5-HD[4]. In ischemic preconditioning activated p38 MAPK causes opening of K_{ATP} channels on the inner mitochondrial membrane by mechanisms that may involve HSP27 and actin filaments.[4]

The effectors that are responsible for the protective events in late preconditioning have not yet been identified but several strong candidates have been proposed which include K_{ATP} channels, NOS, cyclooxygenase-2 (COX-2), aldose reductase and antioxidant enzymes.[62,55] In late preconditioning against infarction a diverse range of stimuli including heat stress, ischemia and adenosine receptor activation all converge on K_{ATP} channels suggesting that these channels may constitute a common downstream mechanism of protection. By contrast K_{ATP} channel activity is not required for ischemia induced late preconditioning against stunning.[60]

Opening of K_{ATP} channels is critical for the late phase of protection against infarction induced by the endotoxin derivative monophosphphoryl lipid A (MLA), or by heat stress, ischemia, adenosine receptor agonists or delta-opioid receptor agonists.[7] Because of the diverse range of stimuli that converge on K_{ATP} channels it has been proposed that the channels may be common distal effectors of late preconditioning. The identity of the K_{ATP} channels involved in late

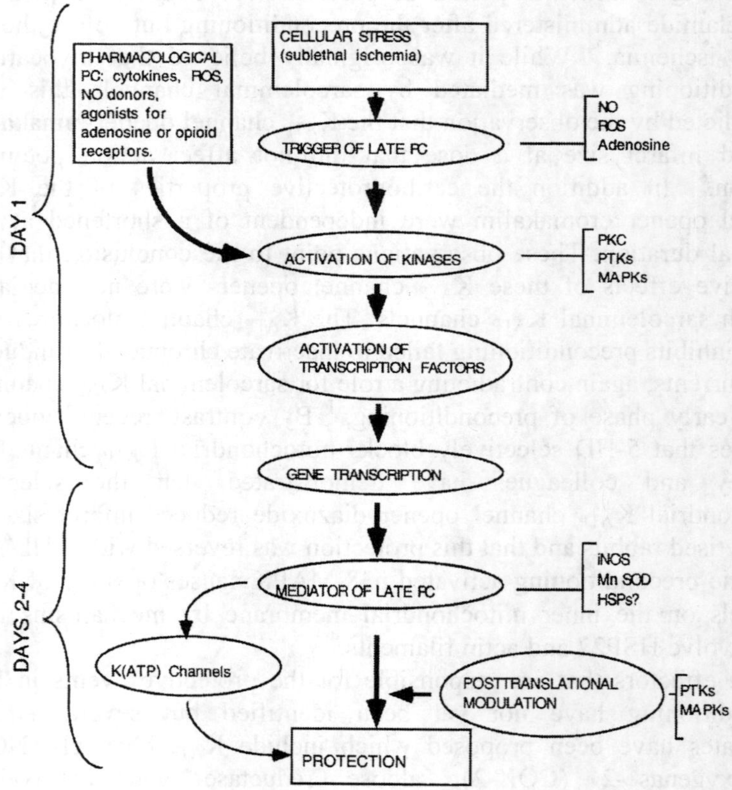

Fig. 9.1 Schematic representation of the cellular mechanisms underlying late preconditioning showing triggers, regulatory factors including kinases and transcription factors and the mediators including iNOS, HSPs, MAPKs and K(ATP) channels (PC: preconditioning) (after Bolli, 2000).

myocardial preconditioning (i.e. sarcolemmal or mitochondrial) has yet to be determined.

5. NITRIC OXIDE AND REACTIVE OXYGEN SPECIES

It was first demonstrated by Bolli and colleagues (1997)[7,8] that administration of N-nitro-L-arginine (L-NA) an inhibitor of all 3 NO synthase isoenzymes neuronal (nNOS), endothelial (eNOS) and inducible (iNOS), would block delayed protection against stunning which thus implicated NO generation as a trigger in this process. In particular eNOS is likely to be the source of enhanced NO in preconditioning, as the non-specific NOS inhibitor l-NA will block cardioprotection, whereas the iNOS specific inhibitor aminoguanidine does not prevent late preconditioning.[8]

In a mouse model of myocardial infarction it was found that late preconditioning is associated with upregulating iNOS in cardiac myocytes and that a knockout of the iNOS gene blocks the delayed protection against infarction.[26] Studies on iNOS knock out mice have also implicated iNOS as an effector for the delayed infarct size limitation induced by NO donors, delta opioid receptors, exercise, pretreatment with endotoxin derivatives or by diazoxide.[26,7]

One further potential trigger for late preconditioning is ROS generation. Administration of a combination of antioxidants during the initial ischemic episode abolished the protective response against stunning, infarction, arrhythmias and coronary endothelial injury. This treatment also blocks heat stress and exercise induced late preconditioning against infarction.[63]

6. PROTEIN KINASE C AND OTHER SIGNALING KINASES

There is good evidence that in early myocardial preconditioning, receptor triggers at the level of the membrane and preconditioning pathways involving the release of oxygen radicals, all converge on a common downstream pathway- the protein kinase C (PKC) pathway. PKC inhibition in rabbit or rat heart has been shown to prevent the protection afforded by ischemic preconditioning and administration of phorbol 12-myristate 13-acetate (PMA) the PKC activator results in protection of human cardiac myocytes.[66,43] A prerequisite for PKC activation is its translocation to the particulate fraction a process that in rabbit models of myocardial preconditioning is found in particular with a specific isoform of PKC – PKC-epsilon.[43,48] The change in localisation of PKC isoforms within the cell is intrinsic to the process of activation of the enzyme and is involved in mediating a switch to alternative anchoring proteins in the new sub-cellular compartment.

Late preconditioning is mediated by a cascade of signaling events resulting in transcriptional upregulation of cardioprotective genes. In

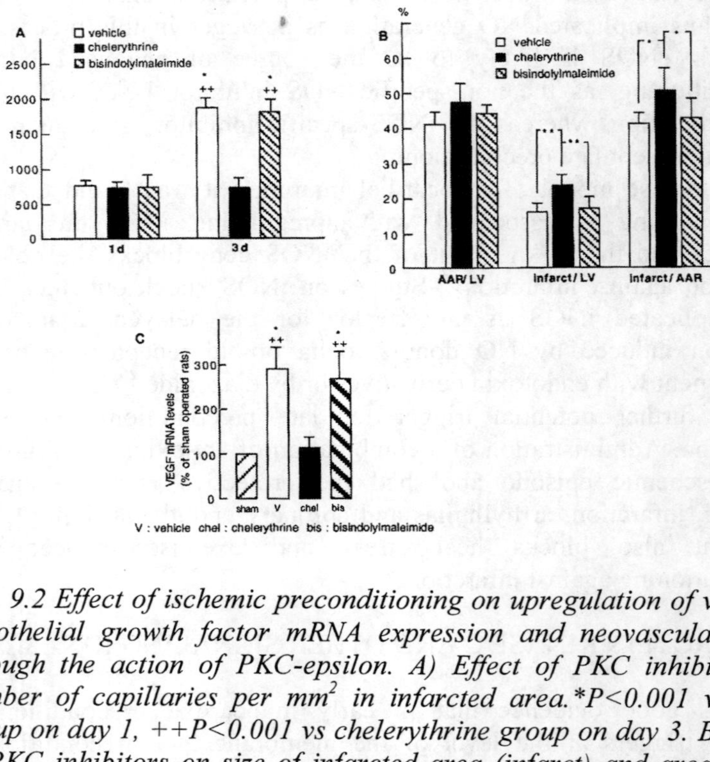

*Fig. 9.2 Effect of ischemic preconditioning on upregulation of vascular endothelial growth factor mRNA expression and neovascularisation through the action of PKC-epsilon. A) Effect of PKC inhibitors on number of capillaries per mm^2 in infarcted area. *P<0.001 vs same group on day 1, ++P<0.001 vs chelerythrine group on day 3. B) Effect of PKC inhibitors on size of infarcted area (infarct) and area at risk (AAR) at 3 days after infarction, *P<0.05, **P<0.01. C) Effect of PKC inhibitors on VEGF mRNA levels in infarcted area. Messenger RNA levels expressed as percentage of mean of sham group (100%). *P<0.0001 vs sham group, ++P<0.0001 vs chelerythrine group. (chel.: inhibits PKC epsilon; bis.: inhibits other PKC isoforms but not PKC epsilon) (after Kawata et al., 2001).*

particular PKC is activated after late ischemic preconditioning with an isoform specific effect on PKC-epsilon.[46] Activation of PKC-epsilon in delayed protection is caused by NO and can be mimicked by NO donors.[46] It has been proposed that NO derived reactive species (peroxynitrite and/or reactive oxygen species) may activate PKC-epsilon either through protein oxidation or through activation of phospholipases. In the early phase and possibly the late phase of preconditioning in rat heart, PKC delta or epsilon mediates protection through membrane translocation. It has also recently been confirmed in the in vivo model of myocardial infarction in rat heart that PKC-epsilon will translocate to the nucleus. Under these circumstances PKC-epsilon enhances expression of the VEGF gene perhaps through message stabilization and thereby elicits angiogenesis and late ischemic preconditioning.[31]

Late ischemic preconditioning selectively activates the protein tyrosine kinase (PTK) family members src and Lck but not epidermal growth factor (EGF) receptor PTKs. Ischemia induced late preconditioning in mice has since been shown to be mediated by phosphorylation of Lck by PKC.[49] While the nature of the downstream effectors that are activated by PKC-epsilon during the development of late preconditioning is unresolved there is now evidence from studies on conscious rabbits for the involvement of p44/p42 MAPKs and JNKs in this process. Unlike these MAPK family members, p38 MAPK is activated initially only for a brief period and its activation is not known to be PKC dependent. The recent finding that p38 MAPK is increased 24h after administration of CCPA, an adenosine A1 selective agonist, suggests a role of this kinase in signaling, firstly during the development of late preconditioning and later as a mediator in aspects of the protective process.

7. ISCHEMIC PRECONDITIONING IN BRAIN

The discovery that preconditioning applies to brain as well as heart and other tissues has inevitably elicited tremendous interest because it reveals that there are mechanisms for achieving tolerance to anoxia in mammalian brain albeit for a much shorter period of time than in the turtle. For the most part the effects of preconditioning in the brain may be considered analogous to the phenomenon of late preconditioning in the heart. The processes that underlie ischemic preconditioning in brain are revealing in terms of targets for pharmacological treatments and novel gene therapy interventions for stroke.

Delayed ischemic tolerance in the brain takes 2 or more days to become maximal and lasts for approximately 7 days.[39,38,13] Preconditioning in brain parallels in a number of ways the process of late myocardial ischemic preconditioning, in particular because of the delayed nature of this tolerance in brain and by virtue of the common molecular pathways involved. Brain ischemic tolerance is mediated to a large extent through activation of N-methyl-D-aspartate (NMDA) glutamate receptors and through increases in intracellular Ca^{2+}.[23] Like late myocardial preconditioning delayed ischemic tolerance in brain may involve activation of adenosine receptors and K_{ATP} channels, or generation of nitric oxide, and is associated with changes in protein phosphorylation.[28,20,54]

Recent evidence on an oxygen-glucose deprivation cell culture model of cerebral ischemia has indicated that preconditioning may involve a signaling cascade that is mediated by activation of the NMDA receptor, Ca^{2+} influx and generation of NO leading to ERK (Ras pathway) mediated tolerance.[21] Previous studies in heart had implicated an alternative pathway involving NO that is PKC dependent.[46] The inducible form of NOS is critical for delayed preconditioning in heart but not in brain. By contrast endothelial NOS is critical for ischemic preconditioning of immature brain[20] and nNOS plays a prominent role in ischemic preconditioning in mature neurones.[21] Studies in heart had implicated p44/p42 MAPKs also known as ERKs in an alternative pathway in preconditioning that is PKC dependent.[47] While ERKs have been proposed as downstream mediators of PKC signaling in myocardium their functional role in cardiac preconditioning has not as yet been definitively characterized.[57] Late ischemic preconditioning may therefore employ different transduction pathways such as PKC signaling possibly through the ERK cascade as seen in heart or Ras dependent pathways as found in neuronal oxygen-glucose deprivation (OGD) tolerance. NO plays a role in both the PKC and Ras dependent signaling pathways. The importance of Ras/ERK mediated tolerance demonstrated in the neuronal model is supported by in vivo observations of phosphorylation of MEK and ERK in the hippocampus in global cerebral ischemic preconditioning.[54] Phosphorylation of these two enzymes after sublethal ischemia was found to correlate with the neuroprotection induced by preconditioning. It is possible that this protection could be elicited by transcriptional activation of neuroprotective genes through phosphorylation of transcription factor targets such as c-fos, ATF-2, cMyc, c-Myb, STAT3 or Ets2.[31]

Like the heart, activation of K_{ATP} channels has been found to protect the brain against the damaging effects of ischemia.[53] This protection may occur through inhibition of synaptic glutamate release or through post synaptic hyperpolarisation that may also result in a decrease in glutamate excitotoxicity. In cerebral ischemic preconditioning in rats Lazdunski and colleagues recently demonstrated that the endogenous protective mechanisms in repetitive brain ischemia could be triggered by activation of adenosine A1 receptors which open K_{ATP} channels which in turn inhibit synaptic glutamate release.[14,16] K_{ATP} channels can also be activated by a range of hormones, neurotransmitters and signaling pathways. It is possible therefore that in cerebral preconditioning, in addition to adenosine, neurotransmitters or peptides may be released that are capable of activating K_{ATP} channels and eliciting a protective effect.[14,33,15]

The multifunctional cytokine, TNF-alpha, plays a role in mediating many of the damaging effects of cerebral ischemia.[39,56] The sphingolipid ceramide is a second messenger in many of the signaling pathways initiated by TNF-alpha receptor occupancy.[19,34] Although many studies have implicated ceramide in the induction of apoptosis it has also been found that ceramide can be cytoprotective for hippocampal neurons. In an in vitro model of neuronal hypoxic preconditioning it was recently demonstrated that protection against hypoxia and oxygen-glucose deprivation injury (Fig. 9.3) was mediated by ceramide synthesis that was triggered by TNF-alpha.[40] One possible mechanism for TNF-alpha and ceramide induced cytoprotection would involve a role for activated NF-kappaB in alterating intracellular Ca^{2+} concentration and outward K^+ currents or in inducing antioxidants such as superoxide dismutase.[5,17,40]

While VEGF gene expression is enhanced in brain following ischemia or hypoxia it is unknown whether VEGF mediates late preconditioning in brain as it does in heart. In heart protein kinase C induced stabilization of VEGF mRNA may contribute to late preconditioning. In brain VEGF protein is induced within 1 h after a 90 min transient middle cerebral artery occlusion.[27] A number of related studies have confirmed elevations of VEGF levels that persist beyond the time of ischemia into the reperfusion phase with a drop off to background levels by 12 h.[9.35,37] By contrast in a rat model of brief

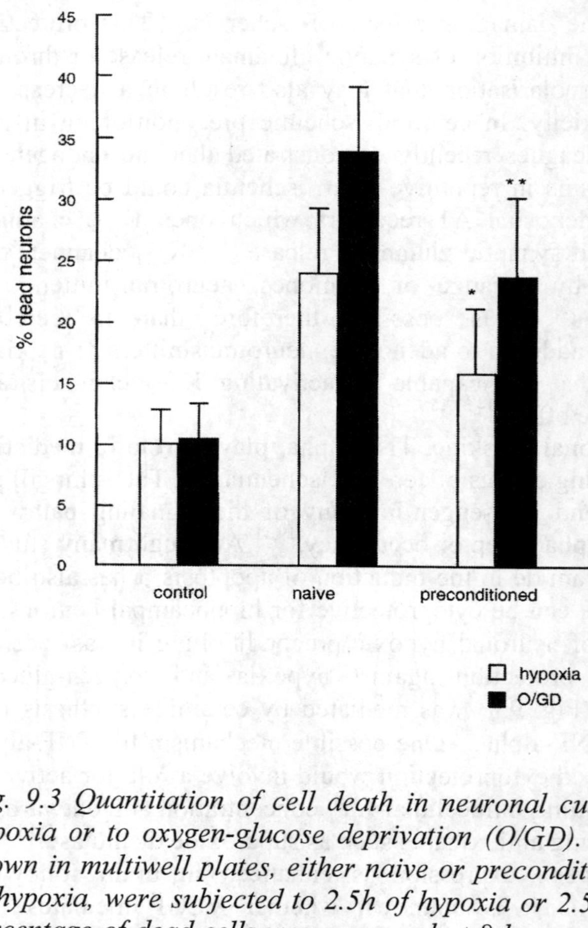

*Fig. 9.3 Quantitation of cell death in neuronal cultures subjected to hypoxia or to oxygen-glucose deprivation (O/GD). Neuronal cultures grown in multiwell plates, either naive or preconditioned with 20 min. of hypoxia, were subjected to 2.5h of hypoxia or 2.5 h of (O/GD), and percentage of dead cells was measured at 8 hours after reoxygenation by means of ethidium assay. Each bar represents mean ± SD of 4 experiments. *Significant difference from hypoxia-induced injury, **significant difference from O/GD induced injury.(after Gonzalez-Zulueta et al., 2000).*

reversible global ischemia VEGF levels were increased 12 h after the ischemic episode and stayed elevated beyond 48 h returning to baseline after 7 days.[45] This long term elevation of VEGF expression may be a result of transcriptional activation of the VEGF gene through HIF-1-alpha binding to the VEGF promoter or potentially through VEGF message stabilization.

8. SUMMARY

Ischemic preconditioning is a particularly important phenomenon in heart and brain and is defined by models of tissue protection whereby a brief mild ischemic episode protects a tissue from a later more severe ischemic insult. The molecular pathways that underlie this tissue protection are best understood in heart but, for late preconditioning in particular, the pathways in heart and brain appear very similar. K_{ATP} channels of the plasma membrane or the mitochondria play a role as an effector of tissue protection by responding to a variety of upstream triggers including adenosine receptor activation, NO generation and reactive oxygen species. Protein kinase C isoform specific activation has been implicated as part of a major signal transduction pathway for preconditioning that may in turn signal through a downstream ERK pathway. In the oxyygen glucose deprivation model of neuronal preconditioning activation of the NMDA receptor, calcium influx and generation of NO leads to activation of the an alternative pathway for preconditioning; the Ras pathway. Thus alternative activators including adenosine receptor agonists, growth factors and reactive oxygen species, will elicit protection through downstream signaling pathways such as the Ras or PKC pathways. Recent studies on VEGF expression in brain illustrate how such protective signaling pathways may also translate their lasting functional effects through regulation of synthesis and release of neuroprotective growth factors.

9. REFERENCES

1. Auchampach JA, Grover GJ, Gross GJ. Blockade of ischaemic preconditioning in dogs by the novel ATP dependent potassium channel antagonist sodium 5-hydroxydecanoate. Cardiovasc Res. 1992 Nov;26(11):1054-62.
2. Auchampach JA, Rizvi A, Qiu Y, Tang XL, Maldonado C, Teschner S, Bolli R. Selective activation of A3 adenosine receptors with N6-(3-iodobenzyl)adenosine-5'-N-methyluronamide protects against myocardial stunning and infarction without hemodynamic changes in conscious rabbits. Circ Res. 1997 Jun;80(6):800-9.
3. Baines CP, Goto M, Downey JM. Oxygen radicals released during ischemic preconditioning contribute to cardioprotection in the rabbit myocardium. J Mol Cell Cardiol. 1997 Jan;29(1):207-16.

4. Baines CP, Liu GS, Birincioglu M, Critz SD, Cohen MV, Downey JM. Ischemic preconditioning depends on interaction between mitochondrial KATP channels and actin cytoskeleton. Am J Physiol. 1999 Apr;276(4 Pt 2):H1361-8.
5. Barger SW, Horster D, Furukawa K, Goodman Y, Krieglstein J, Mattson MP Tumor necrosis factors alpha and beta protect neurons against amyloid beta-peptide toxicity: evidence for involvement of a kappa B-binding factor and attenuation of peroxide and Ca2+ accumulation. Proc Natl Acad Sci U S A. 1995 Sep 26;92(20):9328-32.
6. Bolli R The early and late phases of preconditioning against myocardial stunning and the essential role of oxyradicals in the late phase: an overview. Basic Res Cardiol. 1996 Jan-Feb;91(1):57-63.
7. Bolli R.The late phase of preconditioning. Circ Res. 2000 Nov 24;87(11):972-83.
8. Bolli R, Manchikalapudi S, Tang XL, Takano H, Qiu Y, Guo Y, Zhang Q, Jadoon AK. The protective effect of late preconditioning against myocardial stunning in conscious rabbits is mediated by nitric oxide synthase. Evidence that nitric oxide acts both as a trigger and as a mediator of the late phase of ischemic preconditioning. Circ Res. 1997 Dec;81(6):1094-107.
9. Cobbs CS, Chen J, Greenberg DA, Graham SH. Vascular endothelial growth factor expression in transient focal cerebral ischemia in the rat. Neurosci Lett. 1998 Jun 19;249(2-3):79-82.
10. Cohen MV, Baines CP, Downey JM. Ischemic preconditioning: from adenosine receptor of KATP channel. Annu Rev Physiol. 2000;62:79-109.
11. Cohen MV, Yang XM, Downey JM. Conscious rabbits become tolerant to multiple episodes of ischemic preconditioning. Circ Res. 1994 May;74(5):998-1004.
12. Currie RW, Karmazyn M, Kloc M, Mailer K.Heat-shock response is associated with enhanced postischemic ventricular recovery. Circ Res. 1988 Sep;63(3):543-9.
13. Dawson VL, Dawson TM. Neuronal ischaemic preconditioning. Trends Pharmacol Sci. 2000 Nov;21(11):423-4.
14. de Weille JR, Schmid-Antomarchi H, Fosset M, Lazdunski M. Regulation of ATP-sensitive K+ channels in insulinoma cells: activation by somatostatin and protein kinase C and the role of cAMP.Proc Natl Acad Sci U S A 1989 Apr;86(8):2971-5.
15. Dunne MJ, Bullett MJ, Li GD, Wollheim CB, Petersen OH. Galanin activates nucleotide-dependent K+ channels in insulin-secreting cells via a pertussis toxin-sensitive G-protein. EMBO J 1989 Feb;8(2):413-20
16. Fredholm BB, Dunwiddie TV.How does adenosine inhibit transmitter release? Trends Pharmacol Sci 1988 Apr;9(4):130-4
17. Furukawa K, Mattson MP. The transcription factor NF-kappaB mediates increases in calcium currents and decreases in NMDA- and AMPA/kainate-induced currents induced by tumor necrosis factor-alpha in hippocampal neurons. J Neurochem. 1998 May;70(5):1876-86.
18. Garlid KD, Paucek P, Yarov-Yarovoy V, Murray HN, Darbenzio RB, D'Alonzo AJ, Lodge NJ, Smith MA, Grover GJ. Cardioprotective effect of diazoxide and its interaction with mitochondrial ATP-sensitive K+ channels. Possible mechanism of cardioprotection. Circ Res. 1997 Dec;81(6):1072-82.
19. Ghosh S, Strum JC, Bell RM. Lipid biochemistry: functions of glycerolipids and sphingolipids in cellular signaling. FASEB J. 1997 Jan;11(1):45-50
20. Gidday JM, Shah AR, Maceren RG, Wang Q, Pelligrino DA, Holtzman DM, Park TS.Nitric oxide mediates cerebral ischemic tolerance in a neonatal rat model of hypoxic preconditioning. J Cereb Blood Flow Metab 1999 Mar;19(3):331-40
21. Gonzalez-Zulueta M, Feldman AB, Klesse LJ, Kalb RG, Dillman JF, Parada LF, Dawson TM, Dawson VL. Requirement for nitric oxide activation of p21(ras)/extracellular regulated kinase in neuronal ischemic preconditioning. Proc Natl Acad Sci U S A 2000 Jan 4;97(1):436-41
22. Goodman Y, Mattson MP. Ceramide protects hippocampal neurons against excitotoxic and oxidative insults, and amyloid beta-peptide toxicity. J Neurochem. 1996 Feb;66(2):869-72.
23. Grabb MC, Choi DW. Ischemic tolerance in murine cortical cell culture: critical role for NMDA receptors. J Neurosci 1999 Mar 1;19(5):1657-62
24. Gross GJ, Auchampach JA Blockade of ATP-sensitive potassium channels prevents myocardial preconditioning in dogs. Circ Res. 1992 Feb;70(2):223-33.

25. Grover GJ, D'Alonzo AJ, Hess T, Sleph PG, Darbenzio RB.Glyburide-reversible cardioprotective effect of BMS-180448 is independent of action potential shortening. Cardiovasc Res. 1995 Nov;30(5):731-8.
26. Guo Y, Jones WK, Xuan YT, Tang XL, Bao W, Wu WJ, Han H, Laubach VE, Ping P, Yang Z, Qiu Y, Bolli R.The late phase of ischemic preconditioning is abrogated by targeted disruption of the inducible NO synthase gene. Proc Natl Acad Sci U S A. 1999 Sep 28;96(20):11507-12.
27. Hayashi T, Abe K, Suzuki H, Itoyama Y. Rapid induction of vascular endothelial growth factor gene expression after transient middle cerebral artery occlusion in rats. Stroke. 1997 Oct;28(10):2039-44.
28. Heurteaux C, Lauritzen I, Widmann C, Lazdunski M. Essential role of adenosine, adenosine A1 receptors, and ATP-sensitive K+ channels in cerebral ischemic preconditioning. Proc Natl Acad Sci U S A 1995 May 9;92(10):4666-70
29. Hide EJ, Thiemermann C. Limitation of myocardial infarct size in the rabbit by ischaemic preconditioning is abolished by sodium 5-hydroxydecanoate. Cardiovasc Res. 1996 Jun;31(6):941-6.
30. Kaszala K, Vegh A, Papp JG, Parratt JR. Time course of the protection against ischaemia and reperfusion-induced ventricular arrhythmias resulting from brief periods of cardiac pacing. J Mol Cell Cardiol. 1996 Oct;28(10):2085-95.
31. Kawata H, Yoshida K, Kawamoto A, Kurioka H, Takase E, Sasaki Y, Hatanaka K, Kobayashi M, Ueyama T, Hashimoto T, Dohi K. Ischemic preconditioning upregulates vascular endothelial growth factor mRNA expression and neovascularization via nuclear translocation of protein kinase C epsilon in the rat ischemic myocardium. Circ Res. 2001 Apr 13;88(7):696-704.
32. Kim SJ, Kahn CR. Insulin regulation of mitogen-activated protein kinase kinase (MEK), mitogen-activated protein kinase and casein kinase in the cell nucleus: a possible role in the regulation of gene expression. Biochem J 1997 May 1;323 (Pt 3):621-7
33. Kirsch GE, Codina J, Birnbaumer L, Brown AM. Coupling of ATP-sensitive K+ channels to A1 receptors by G proteins in rat ventricular myocytes. Am J Physiol 1990 Sep;259(3 Pt 2):H820-6
34. Kolesnick R, Golde DW. The sphingomyelin pathway in tumor necrosis factor and interleukin-1 signaling. Cell. 1994 May 6;77(3):325-8.
35. Kovacs Z, Ikezaki K, Samoto K, Inamura T, Fukui M.VEGF and flt. Expression time kinetics in rat brain infarct. Stroke. 1996 Oct;27(10):1865-72; discussion 1872-3.
36. Kuzuya T, Hoshida S, Yamashita N, Fuji H, Oe H, Hori M, Kamada T, Tada M.Delayed effects of sublethal ischemia on the acquisition of tolerance to ischemia. Circ Res. 1993 Jun;72(6):1293-9.
37. Lennmyr F, Ata KA, Funa K, Olsson Y, Terent A. Expression of vascular endothelial growth factor (VEGF) and its receptors (Flt-1 and Flk-1) following permanent and transient occlusion of the middle cerebral artery in the rat. J Neuropathol Exp Neurol. 1998 Sep;57(9):874-82.
38. Lipton P.Ischemic cell death in brain neurons. Physiol Rev 1999 Oct;79(4):1431-568
39. Liu J, Ginis I, Spatz M, Hallenbeck JM. Hypoxic preconditioning protects cultured neurons against hypoxic stress via TNF-alpha and ceramide. Am J Physiol Cell Physiol 2000 Jan;278(1):C144-53
40. Mattson MP, Goodman Y, Luo H, Fu W, Furukawa K. Activation of NF-kappaB protects hippocampal neurons against oxidative stress-induced apoptosis: evidence for induction of manganese superoxide dismutase and suppression of peroxynitrite production and protein tyrosine nitration. J Neurosci Res. 1997 Sep 15;49(6):681-97.
41. McCullough JR, Normandin DE, Conder ML, Sleph PG, Dzwonczyk S, Grover GJ. Specific block of the anti-ischemic actions of cromakalim by sodium 5-hydroxydecanoate. Circ Res. 1991 Oct;69(4):949-58.
42. Miki T, Cohen MV, Downey JM. Opioid receptor contributes to ischemic preconditioning through protein kinase C activation in rabbits. Mol Cell Biochem. 1998 Sep;186(1-2):3-12.

43. Mitchell MB, Meng X, Ao L, Brown JM, Harken AH, Banerjee A. Preconditioning of isolated rat heart is mediated by protein kinase C.Circ Res. 1995 Jan;76(1):73-81.
44. Parratt JR. Protection of the heart by ischaemic preconditioning: mechanisms and possibilities for pharmacological exploitation. Trends Pharmacol Sci. 1994 Jan;15(1):19-25.
45. Pichiule P, Chavez JC, Xu K, LaManna JC. Vascular endothelial growth factor upregulation in transient global ischemia induced by cardiac arrest and resuscitation in rat brain. Brain Res Mol Brain Res. 1999 Dec 10;74(1-2):83-90.
46. Ping P, Takano H, Zhang J, Tang XL, Qiu Y, Li RC, Banerjee S, Dawn B, Balafonova Z, Bolli R. Isoform-selective activation of protein kinase C by nitric oxide in the heart of conscious rabbits: a signaling mechanism for both nitric oxide-induced and ischemia-induced preconditioning. Circ Res. 1999 Mar 19;84(5):587-604.
47. Ping P, Zhang J, Cao X, Li RC, Kong D, Tang XL, Qiu Y, Manchikalapudi S, Auchampach JA, Black RG, Bolli R. PKC-dependent activation of p44/p42 MAPKs during myocardial ischemia-reperfusion in conscious rabbits. Am J Physiol 1999 May;276(5 Pt 2):H1468-81
48. Ping P, Zhang J, Qiu Y, Tang XL, Manchikalapudi S, Cao X, Bolli R.Ischemic preconditioning induces selective translocation of protein kinase C isoforms epsilon and eta in the heart of conscious rabbits without subcellular redistribution of total protein kinase C activity. Circ Res. 1997 Sep;81(3):404-14.
49. Ping P, Zhang J, Zheng YT, Li RC, Dawn B, Tang XL, Takano H, Balafanova Z, Bolli R. Demonstration of selective protein kinase C-dependent activation of Src and Lck tyrosine kinases during ischemic preconditioning in conscious rabbits. Circ Res. 1999 Sep 17;85(6):542-50.
50. Prosser, C. L., Barr, L. M., Pinc, R. D., and Lauer, C. Y. 1957. Acclimation of goldfish to low concentrations of oxygen. Physiol. Zool. 30, 137-141.
51. Qiu Y, Rizvi A, Tang XL, Manchikalapudi S, Takano H, Jadoon AK, Wu WJ, Bolli R. Nitric oxide triggers late preconditioning against myocardial infarction in conscious rabbits. Am J Physiol. 1997 Dec;273(6 Pt 2):H2931-6.
52. Routley, M. H., Nilsson, G. E., and Renshaw, G. M. C. Exposure to hypoxia primes the respiratory and metabolic responses of the epaulette shark to progressive hypoxia. Comp. Biochem. Physiol. A, in press.
53. Rubino A, Yellon DM. Ischaemic preconditioning of the vasculature: an overlooked phenomenon for protecting the heart? Trends Pharmacol Sci 2000 Jun;21(6):225-30
54. Shamloo M, Rytter A, Wieloch T. Activation of the extracellular signal-regulated protein kinase cascade in the hippocampal CA1 region in a rat model of global cerebral ischemic preconditioning. Neuroscience 1999;93(1):81-8
55. Shinmura K, Tang XL, Wang Y, Xuan YT, Liu SQ, Takano H, Bhatnagar A, Bolli R. Cyclooxygenase-2 mediates the cardioprotective effects of the late phase of ischemic preconditioning in conscious rabbits. Proc Natl Acad Sci U S A. 2000 Aug 29;97(18):10197-202.
56. Shohami E, Ginis I, Hallenbeck JM. Dual role of tumor necrosis factor alpha in brain injury. Cytokine Growth Factor Rev. 1999 Jun;10(2):119-30.
57. Strohm C, Barancik T, Bruhl ML, Kilian SA, Schaper W.Inhibition of the ER-kinase cascade by PD98059 and UO126 counteracts ischemic preconditioning in pig myocardium. J Cardiovasc Pharmacol 2000 Aug;36(2):218-29
58. Takano H, Tang XL, Qiu Y, Guo Y, French BA, Bolli R. Nitric oxide donors induce late preconditioning against myocardial stunning and infarction in conscious rabbits via an antioxidant-sensitive mechanism. Circ Res. 1998 Jul 13;83(1):73-84.
59. Takano H, Tang XL, Bolli R. Differential role of K(ATP) channels in late preconditioning against myocardial stunning and infarction in rabbits. Am J Physiol Heart Circ Physiol. 2000 Nov;279(5):H2350-9.
60. Takano H, Bolli R, Black RG Jr, Kodani E, Tang XL, Yang Z, Bhattacharya S, Auchampach JA. A(1) or A(3) adenosine receptors induce late preconditioning against infarction in conscious rabbits by different mechanisms. Circ Res. 2001 Mar 16;88(5):520-8.

61. Thornton JD, Liu GS, Olsson RA, Downey JM. Intravenous pretreatment with A1-selective adenosine analogues protects the heart against infarction. Circulation. 1992 Feb;85(2):659-65.
62. Xuan YT, Tang XL, Qiu Y, Banerjee S, Takano H, Han H, Bolli R.Biphasic response of cardiac NO synthase isoforms to ischemic preconditioning in conscious rabbits. Am J Physiol Heart Circ Physiol. 2000 Nov;279(5):H2360-71.
63. Yamashita N, Hoshida S, Otsu K, Asahi M, Kuzuya T, Hori M. Exercise provides direct biphasic cardioprotection via manganese superoxide dismutase activation. J Exp Med. 1999 Jun 7;189(11):1699-706.
64. Yang XM, Baxter GF, Heads RJ, Yellon DM, Downey JM, Cohen MV. Infarct limitation of the second window of protection in a conscious rabbit model. Cardiovasc Res. 1996 May;31(5):777-83.
65. Yao Z, Gross GJ. Effects of the KATP channel opener bimakalim on coronary blood flow, monophasic action potential duration, and infarct size in dogs. Circulation. 1994 Apr;89(4):1769-75.
66. Ytrehus K, Liu Y, Downey JM. Preconditioning protects ischemic rabbit heart by protein kinase C activation. Am J Physiol. 1994 Mar;266(3 Pt 2):H1145-52.

CHAPTER 10

CLINICAL PERSPECTIVES

In this final chapter we ask if the rather esoteric anoxia tolerant animals really provide any useful perspectives for understanding why the mammalian brain fails so rapidly in anoxia or are they just exemplars, albeit fascinating, of a limit in vertebrate evolution? Will such studies help to identify the most vulnerable process(es), the Achilles heel, that result in neuronal death in mammals? It is possible that even man's tolerance to hypoxia may be enhanced if we can identify adaptive mechanisms of other species that can be made applicable to human biology.[17,18,26] The basic argument for believing that such comparative investigations may provide useful clinical insights is that in the energy deprived mammal brain the degenerative events are so rapid and complex that it has proven extremely difficult to sort out cause and effect. In anoxia tolerant species on the other hand, we have unique animal models where the control situation in anoxia is survival instead of death. There appears to be wide agreement that identifying the key processes that are protected in anoxic tolerant animals, as against those that are allowed to decline, may indicate promising targets for successful clinical intervention.[27] The main targets are the regulation of the controlled metabolic suppression and the mechanisms of preservation of cell structure and membrane functions and integrity under conditions of severely reduced energy supplies.[18] Such studies have potentially far-reaching clinical implications including the prevention of secondary brain damage following brain trauma and ischemia as well as induction of a state of neuroprotection during surgical procedures that require temporary circulatory arrest.[18]

The Brain Without Oxygen. Third Edition. Peter L. Lutz, Göran E. Nilsson and Howard M. Prentice.

But this chapter is not in any way intended to be a survey of the vast field of putative remedial and protective interventions in hypoxia/ischemia – it is a selection of a few examples that illustrate the themes revealed in this book.

As we have seen the basic strategies of the anoxia tolerant brain are; first of all to prevent anoxic depolarization and secondly to protect against the consequences of anoxic depolarization. To some extent both strategies are used in the mammal.

1. ANOXIC DEPOLARIZATION DELAYED

The principal strategy of the anoxia tolerant brain is to avoid or delay depolarization rather than combat the consequences. As we have seen, from fish to mammals, a critical stage is reached during anoxia when the slowly rising extracellular K^+ reaches a threshold (phase 2) that triggers depolarization (phase 3). Anoxia tolerant species greatly delay the time taken to reach the depolarizing threshold by depressing the rate of increase in extracellular K^+ primarily by decreasing the activity of the anoxic brain through a variety of mechanisms.

Since in the mammalian brain the anoxia induced increase in extracellular K^+ eventually leads to depolarization and neuronal death there is a substantial interest in finding methods to reduce K^+ efflux. However, the major contributory pathways through which K^+ exits the anoxic cell are unknown, and may differ in different species and in different regions of the brain. For example in the substantia nigra of the rat anoxic K^+ efflux is effectively blocked by sulfonylureas blockers of neuronal ATP dependent K^+ channels.[2] But in the brain stem, blocking the ATP dependent K^+ channels with glibenclamide only reduces K^+ efflux by 43%[30] and in the hippocampus anoxic K^+ efflux is insensitive to any type of known blockers for ATP sensitive, Ca^{2+} sensitive and voltage sensitive K^+ channels.[50] Indeed, the turtle brain appears to have a very much lower density of ATP sensitive K^+ channels compared to the mammal.[71]

A second approach is to prevent the sodium influx that occurs during phase 1 by the pharmacological modulation of voltage dependent Na^+ channels.[65] This early sodium over load can be neurotoxic because excessive intracellular Na^+ can inhibit anaerobic glycolysis, result in energy failure and cell depolarization and the Na^+ influx is associated with swelling and bleb formation.[65] There is evidence that blocking the voltage dependent Na^+ channel with TTX protects against ischemia/anoxia induced neuronal damage.[4] TTX acts to delay the ischemia induced sodium influx, which has the effect of attenuating the ECS shrinkage and postponing anoxic depolarization (Fig 10.1). But perhaps the most general lesson from comparative studies on brain anoxic survival is the primacy of depressing the brains metabolic demand.

The simplest way to reduce metabolic demand is to lower temperature, and as Fig 7.1 in Chapter 7 illustrates there is a striking correlation between anoxia survival time and body temperature throughout the vertebrates. However, as we have discussed, the only mammals that achieve deep hypothermia, with its concomitant enhanced tolerance of hypoxia, are those that hibernate. Normally a severe reduction in temperature is not tolerated well. In adult mammals for example, holding the body temperature at 29°C for 48 h can produce detrimental neurological effects.[61] On the other hand, there is much evidence that even a slight reduction in brain temperature confers substantial benefits to the ischemic/anoxic brain. For example, Ginsberg et al.[20], showed that even a few °C reduction in brain temperature could substantially effect the extent of ischemic injury. By lowering brain temperature of the rat from 36 °C to 30 °C they completely eliminated striatal pathology and greatly reduced hippocampal injury. Hypothermia may also produce a substantial amelioration of neuronal damage during the early postischemic period.[21] Hypothermia not only prolongs the delay to hypoxic depolarization in the gerbil neocortical slice, but by cooling the slice by as little as 2.5 °C the recovery of synaptic responses following 20 min of hypoxic depolarization increased from almost zero (no recovery) to around 27% of the predepolarization maximum evoked potential (Fig 10.2).

A reduction in the rate of high-energy phosphate depletion is likely to be one of the major factors in hypothermic protection. But, it appears that at least part of the protective effect is also due to the slowing of excitatory amino acid release at low temperatures. Ginsberg et al[20] showed that while 10-20 min of ischemia produced a 500 fold increase in brain extracellular dopamine and a seven fold rise glutamate at 36 °C, at 30 °C dopamine only increased by 40% and glutamate did not increase at all. In all likeliness, this is a reflection of a poor ability of mammalian neurotransmitter mechanisms to adjust to a reduction in temperature. Still, this maladaptive feature may prove to be very useful in clinical attempts to provide protection against anoxic damage. But, in practice, while hypothermia has been accepted in human open-heart surgery, it has not found favor in neurosurgery for fear that even mild localized brain hypothermia may result in functional impairment.

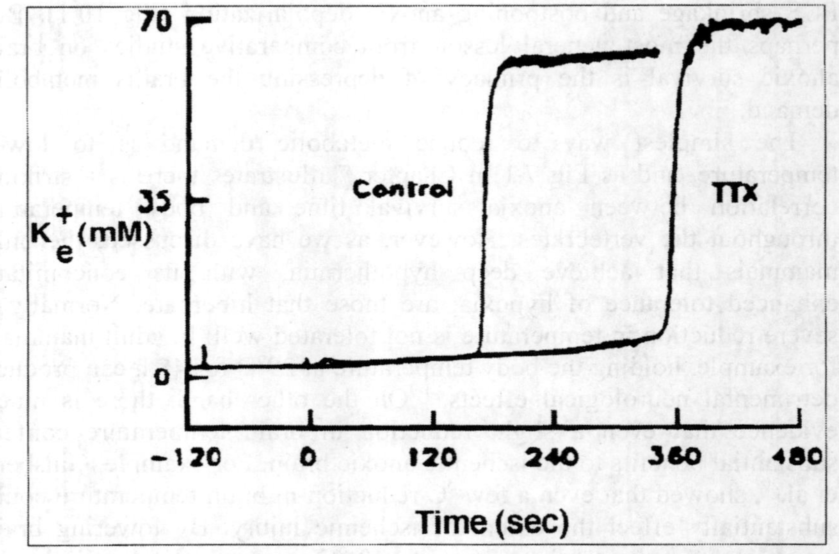

Fig.10.1 Effect of TTX on ischemia induced changes on extracellular K^+ in the rat brain·From Xie et al.[72]

Hypothermia can also moderate or prevent the ischemia related decrease in protein kinase C activity (PKC), an important factor in the pathophysiology of ischemic damage.[6] However, it has also been suggested that the temperature effect is primarily through its influence on membrane permeability to ions.[32] Katsura et al[32] found that lowering the body temperature of rats to 34 °C reduced the initial (phase 1) rate of increase in extracellular K^+ during anoxia, thereby delaying the terminal depolarization. However, the threshold concentration of extracellular K^+ that triggers the phase 2 depolarization was the same for hypothermic, normothermic and hyperthermic (40 °C) animals (10-15 mM). Temperature control of the degree of activation of K^+ conductances was thought to the key factor involved.

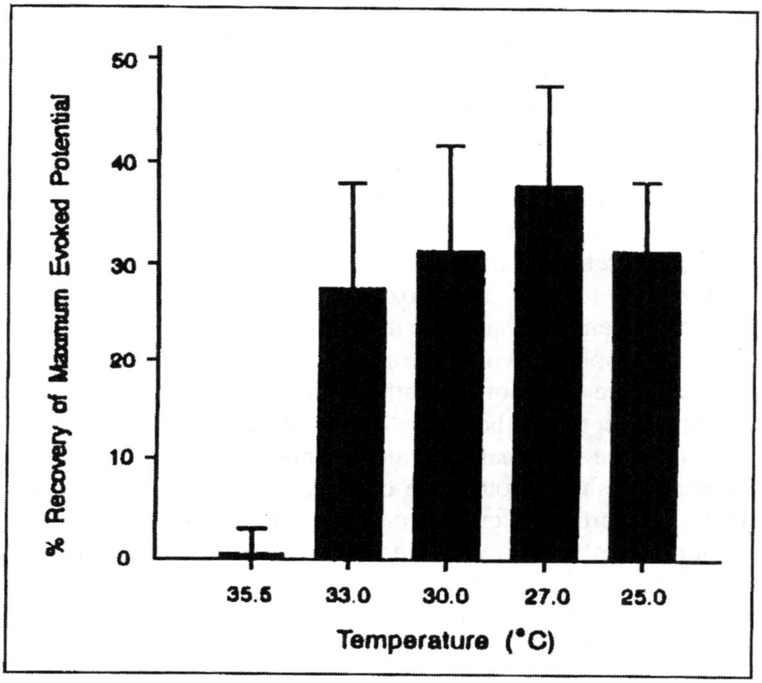

Fig. 10.2 The effect of hypothermia in enhancing the recovery of synaptic responses of gebril neocortical brain slices following 20 min hypoxic depolarization. From Hiramatsu et al.[25]

There are substantial regional differences in the degree to which hypothermia provides protection, linked perhaps to the temperature sensitivities of the processes responsible for selective neuronal vulnerabilities to ischemia/anoxia.[52] For example, in rats that were subjected to 15 min forebrain ischemia, lowering the brain temperature to 33 °C prevented neuronal necrosis in some but not all brain regions, the effect of temperature in the caudoputamen being much greater than in the CA1 region of the hippocampus.[52]

There is widespread evidence of the effectiveness of hypothermia as a defense against hypoxia in newborns and children, the most dramatic being the reports of children surviving submergence in ice cold water for 40 min without neuropathalogical sequelae.[55] In a clinical study Cordey et al[7] report that for severely hypoxic infants who had failed to respond after 5 min of standard resuscitative measures submergence in chilled water (10-15 °C) for 3-15 min facilitated short term recovery and also provided protection against long-term neurological damage. Many experimental studies support this premise. For example, Ikonomidou et al.[29] reported that a relatively small decrease in body temperature markedly reduces hypoxic/ischemic damage in infant rats and improves the survival rate. They found that lowering the body temperature of 10 days old hypoxic rats by 5 °C provided as much attenuation of neurological damage as that produced by administering the NMDA antagonist, MK-801, and that lowering body temperature by only 2.5 °C together with pretreatment with MK-801 provided total protection.

Both anoxia tolerant and anoxia intolerant species reduce neuronal metabolic demand by suppressing brain electrical activity. This is achieved by stopping synaptic transmission using mechanisms that, to some extent, are common to both groups. However, there may be an important exception: the presynaptic hyperpolarization commonly employed by the mammalian brain to depress synaptic transmission is partly accomplished through the opening of ATP or Ca^{2+} activated K^+ channels. The turtle and crucian carp appear to avoid such a mechanism during long term anoxia, which points out an important limitation with this solution: the increased K^+ conductance will produce a gradual rise in extracellular K^+ which if prolonged will result in neuronal depolarization. Thus, unless ATP levels can be restored in a very short run, this strategy would appear to be self defeating and as such will not be used by anoxia-tolerant species.

As discussed in chapter 7, adenosine may "set the stage" for neuronal down regulation and how this is achieved may be best revealed in the anoxia-tolerant animal. This is also the case in mammals where extracellular adenosine appears to serve as a critical "retaliatory metabolite" during hypoxia. There is now overwhelming evidence that activation of adenosine A_1 receptors can reduce ischemic and hypoxic damage in some brain regions and endogenous adenosine even appears to provide important neuronal protection during prolonged moderate hypoxia in rats (40 min episodes of 5% O_2 for 3 days.[5] In addition, adenosine appears to participate in long lasting adaptive changes in the mammalian brain, so called ischemic preconditioning. This phenomenon is seen as an increased ability to tolerate ischemia if the brain has been subjected to brief ischemic periods several days earlier. Using specific adenosine A_1 receptor antagonists, Heurteaux et al.[24] were able to block ischemic preconditioning in the rat brain.

In consequence of these findings an intense effort is underway to enhance adenosine's protective effects. Adenosine A_1 receptor agonists are being sought as potential cerebroprotective agents and another approach is to enhance endogenous extracellular levels of adenosine by inhibiting adenosine uptake or breakdown.[64] Indeed there is good experimental evidence that adenosine deaminase inhibitors such as deoxycoformycin and uptake blockers like propentofylline increase ischemia/hypoxia evoked adenosine levels[40,42] and uptake inhibitors are now being tested in humans.[64] However, there are potential drawbacks to this putative therapy. Excessive or prolonged exposure to extracellular adenosine can cause A_1 receptor desensitization with a resultant functional impairment of adenosines actions.[64] Above about 20μmol adenosine may also activate excitatory A_2 receptors which can promote the release of glutamate or excite neurons.[64]

It is very likely that the slow and sustained release of inhibitory neurotransmitters such as GABA is also involved in metabolic depression in anoxia tolerant species, and could play a protective role in mammals.

Accumulating evidence supports this suggestion. Injections of GABA or a GABA-receptor agonist into rats causes the body temperature to decrease by several °C (Sancibrian et al.[49] for review) which would increase hypoxic survival time. Indeed, elevating brain GABA levels in gerbils, through a pharmacological blockade of GABA break down with gamma-vinyl GABA, results in a reduced metabolic rate sufficient to produce a 3 °C decrease in body temperature, and also greatly attenuates ischemic depletion of brain energy stores (ATP, phosphocreatine, glucose and glycogen).[1] Another study has shown that treatment with gamma-vinyl GABA significantly reduces brain damage during cerebral ischemia in gerbils.[53] In fact, there is an increasing number of studies showing that GABA reuptake inhibitors and GABA receptor agonists provide substantial protection against ischemic/anoxic damage (Fig.10.3) in virtually any part of the mammalian CNS (striatum, hippocampus, substantia nigra, thalamus, spinal cord and optic nerve), in some cases also when administered post-ischemia.[15,36,37,52,53,54]

Moreover, there is a report suggesting that the ischemic protection provided by adrenalectomy in gerbils is mediated by an induction of GABA levels [44]. It is of interest to note that in the anoxic turtle brain GABA is released as extracellular adenosine starts to decline, suggesting that it is acting on a brain with already activated adenosine receptors. Indeed, there are mammalian experiments suggesting a synergistic action between adenosine and GABA in protecting CNS white matter against anoxic injury.[14] Presumably, the mechanisms that stimulate GABA release and $GABA_A$ receptor up-regulation during anoxia in the turtle are of potential relevance for the mammalian system.

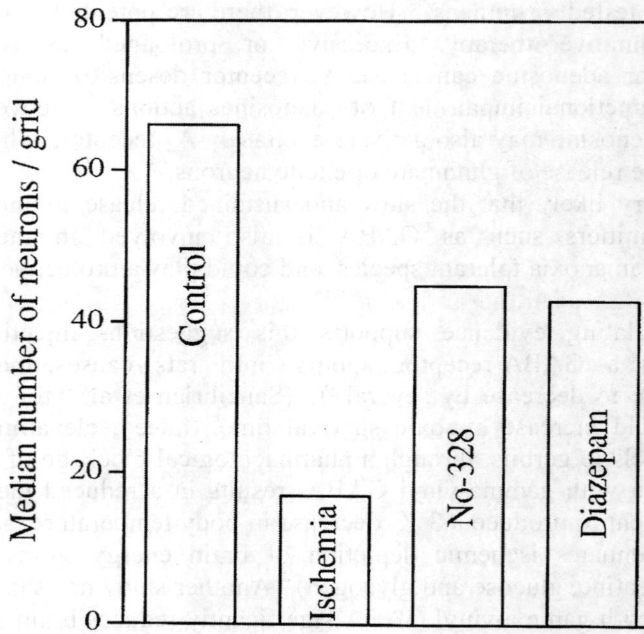

Fig 10.3. Effect of GABAergic stimulation in vivo on survival of hippocampal CA1 pyramidal neurons in rats subjected to 10 min of global ischemia. No-328 is a GABA reuptake blocker, and diazepam is an allosteric stimulator of the $GABA_A$ receptor. The rats were treated repeatedly with the drugs before and after ischemia. Number of pyramidal neurons was counted in a 480 x 96 µm grid. From Johansen et al.[31]

It is probable that much of the protection provided by anesthetics against hypoxia is due to their suppressing effect on cerebral metabolic rate. The long known ability of general anaesthesia to attenuate anoxic/ischemic injury is likely to be related to a metabolic suppression caused by stimulation of the $GABA_A$ receptor - the likely site of action of many general anaesthetics. Moreover, local anesthetics such as lidocaine have been found to depress O_2 consumption and glucose uptake in rat hippocampal slices, probably by reducing Na^+ influx and concomitant Ca^{2+} influx.[35]

After ion transport, protein synthesis and transport are the next most costly activities of the neuron and energy savings might be expected from the curtailment of these activities. Indeed, it has been shown that inhibiting protein synthesis provides some protection from ischemic damage in the rat hippocampus[12], and incubating cerebellar neuron cell cultures with protein synthesis inhibitors results almost doubles the rate of anoxia survival.[12] However, it is very likely that halting protein synthesis and transport for any extended period would compromise the "architecture" of the neuronal network and result in impaired function.

The manipulation of ion channel activities is likely to be an important common strategy for both anoxia tolerant and anoxia intolerant species. In the turtle Ca^{2+} channel activity is reduced during long term anoxia, possibly through the agency of an extracellular protein.[3] As discussed, in the mammal a reduction of Ca^{2+} conductance may be a key factor in blocking neurotransmitter release during the initial hypoxia period. In the turtle an anoxia driven downregulation of Na^+ channels may have the effect of increasing the threshold for the activation of action potentials, and thereby reduce the rate of firing.[41] There is evidence that mammals also have the capacity to modulate Na^+ channel activity.[65] In rats a downregulation of Na^+ channels has been seen in the fetal brain challenged by increased Na^+ flux[7]and chronic hypoxia (6% O_2) produced a substantial increase in Na^+ channel density of the rat carotid body.[48] More directly, it appears that hypoxia reduces cell excitability in human neocortex brain slices by increasing the probability for Na^+ channels to be in the inactive state.[8] After comprehensively reviewing the evidence it has been proposed that pharmacological modulation of voltage-gated Na^+ channels may be an effective strategy against ischemic brain damage.[65]

It is very possible then, that a better understanding of the mechanisms for the downregulation of Na^+ and Ca^{2+} channels could be important for managing a controlled reduction in brain metabolic rate to enhance anoxic/ischemic survival.

After the meager internal energy stores have been depleted, the anoxic brain of all species depends upon anaerobic glycolysis, fueled by an external supply of glucose, as its only source of energy. In the turtle, anoxia produces increases in both the blood flow to the brain and in blood glucose concentrations, greatly enhancing the supply of glycolytic substrate. This appears to be a primary factor in anoxia survival and indeed anoxia endurance is correlated with the amount of the mobilizable glycogen store in the liver. As discussed in chapter 7 there is some controversy concerning the utility or harm produced by increasing glucose supply in mammals. But it appears that the discrepancy may relate somewhat to the circumstances of energy deprivation. In ischemia or stagnant anoxia where the blood flow is halted, glycolysis can result in a harmful accumulation of lactate and H^+ ions and associated poisonous products. In this event increasing glucose supply may be detrimental to recovery.[56] On the other hand, Swanson et al[63] have shown that delivering glucose via a microdialysis probe to ischemic rat caudate nuclei greatly reduced the release of excitotoxic glutamate. In any event, when flow is still maintained, as in incomplete ischemia, or in asphyxia, an increased brain glucose supply appears to be beneficial.[57] Such a premise would be supported from comparative studies and indeed glucose supplementation is currently recommended for oxygen deprived newborns.[67]

2. PROTECTION FROM CONSEQUENCES OF ANOXIC DEPOLARIZATION

The truly anoxia-tolerant brain appears to be less useful for comparison here, as its basic strategy is to avoid depolarization. Maybe the frog brain, with its striking ability so tolerate a considerable period of ATP depletion, and possibly depolarization, will prove to be a better model for comparison. However, some of the putative harmful effectors such as acidosis and hyperglycemia are tolerated extraordinary well by the turtle brain, and the anoxia tolerant brain presumably has a propensity to experience free radical formation when oxygenation is restored.

One of the major targets for anoxia/ischemia therapy is the prevention or alleviation of glutamate neurotoxicity. Particular effort is being put into finding the most effective glutamate antagonists (see Rothman and Olney[48] and Silver et al.[58], for reviews). For example, experiments on cultured hippocampal and neocortical neurons exposed to high levels of glutamate have shown that treatment with NMDA receptor blockers reduce neuronal death even if the blockers are added after glutamate has been removed from the medium, suggesting that the initial exposure to exogenous glutamate stimulated a continued release of endogenous glutamate - the natural neurotransmitter of these neurons.[47]

Another approach is to find the means of inhibiting the release of glutamate.[23] For example, the pyrimidine-derivative BW 1003C87 reduces post-ischemic damage in the cortex of treated rats – an effect that is paralleled by a reduction in glutamate release.[34,47] Adenosine agonists that act presynaptically to reduce glutamate release may similarly improve hypoxia-ischemia survival.

A second major target is to protect against the uncontrolled increase in intracellular Ca^{2+}. And here much interest is being shown in drugs that reduce Ca^{2+} influx. For example, dantroline, which reduces Ca^{2+} release from intracellular stores, provides some degree of protection from glutamate excitotoxicity.[17] In a clinical study dantroline produced significant protection from hypoxia resulting from a case of CO poisoning.[28] Clinical studies with other Ca^{2+} entry blockers such as nimodipine improved stroke recovery in humans.[62] Most interestingly, intravenous administration of the synthetic conopeptide (SNX-111) that selectively blocks N-type voltage gated Ca^{2+} channels provided significant neuroprotection in rats even if administered up to 24 h after the ischemic insult.[66] Such a process may occur naturally in the anoxic turtle brain which appears protected against increasing intracellular Ca^{2+} through a reduction in Ca^{2+} channel activity.[33]

A third target is to find drugs that can function as free radical scavengers, or inhibit the production of free radicals. There has been some success in reducing reperfusion damage through the use of such drugs as superoxide dismutase, dimethylthiourea, which acts as a HO* scavenger, and allopurinol, which is an inhibitor of xanthine oxidase.[57] Interestingly, the turtle brain, unlike that of the guinea pig, is able to maintain high tissue ascorbate levels throughout many hours of anoxia, which suggests that high ascorbate concentrations could be important in protecting the turtle brain against reoxidative injury.[45]

3. NOVEL MOLECULAR TARGETS AND HYPOXIA REGULATED GENE THERAPY

As has been discussed, a number of candidate molecular therapeutic targets for brain and heart ischemia have been identified from mammalian models of tissue protection. For example, adenosine receptors and heat shock proteins have been implicated in models of cerebral and myocardial preconditioning. Components of pathways for adapting to the consequences of cerebral anoxic depolarization are also potential targets, including anti-apoptosis factors such as Bcl-2. New subtractive hybridisation and microarray technologies from anoxia tolerant species are likely to generate a range of novel molecular targets. For example, in the application of suppression subtraction hybridisation and microarray gene expression profiling to the hypoxia tolerant fish Gillichthys mirabilis, 126 distinct hypoxia regulated cDNAs were found, of which 51 were novel expressed sequence tags (ESTs).[22]

Future clinical applications using new gene specific targets will depend upon successful gene transfer, including adequate tissue penetration and sustained gene expression, as well as appropriate temporal and spatial regulation of transgene expression. New viral vectors have proven to be capable of providing effective gene transfer with high level transgene expression.[70] The targeting of expression of a therapeutic protein to the appropriate tissue has been achieved with the incorporation of tissue specific gene promoters into delivery vectors. However, one possible limitation for regulating the time of foreign gene expression may be the requirement for delivery of a drug that elicits the activation of the target gene.[16,19,33] For an effective therapeutic outcome there may be advantages to controlling therapeutic protein production tightly in a manner that is directly linked to an endogenous symptom of the disease process. In ischemic disorders such as stroke and myocardial ischemia, hypoxia may constitute a valuable endogenous signal that could be employed for regulating expression of a therapeutic gene without the necessity for an exogenous drug based activation system.[68]

As we discussed in chapters 3 and 5, mammalian tissues respond to oxygen deprivation by inducing a range of genes involved in anaerobic energy production, erythrogenesis and angiogenesis. The HIF-1 family of transcription factors is central to the transcriptional induction of many of these genes through specific binding to a consensus site or hypoxia response element (HRE) containing the core sequence ACGTG.[69].To direct therapeutic gene expression specifically to ischemic tissues a functional analysis was performed on promoters containing HREs that are incorporated into a minimal muscle specific promoter.[43] Endogenous HRE promoters are activated by hypoxia in a range from three to four fold induction for some glycolytic enzyme genes to greater than 100 fold for the erythropoeitin gene. A rabbit model of ischemia with reperfusion was employed to test the kinetics of induction from a gene transfer vector containing a reporter gene linked to a minimal heart specific promoter containing four tandem HREs.[43] Plasmid DNA containing the promoter-gene cassette was injected into the left ventricle. After 7 days the heart was subjected to a 15 min ischemic episode by a ligation of the coronary artery and luciferase enzyme activity was assayed at intervals after reperfusion. There was a rapid activation of the transgene in response to transient ischemia with almost maximal reporter gene expression after 1 hr of reperfusion and continuing enhanced expression for at least 4 hours after the ischemic episode. Reporter gene expression returned to baseline levels after 8 hours. While these experiments were based on the use of a medium strength HRE that produced a six fold maximal induction more powerful HRE combinations are now capable of eliciting greater than 25 fold transgene induction above baseline.[68] When incorporated into an appropriate viral vector the HRE driven transgene cassette would therefore be capable of eliciting high level expression of a therapeutic product during ischemia with the region of expression restricted to the ischemic zone.

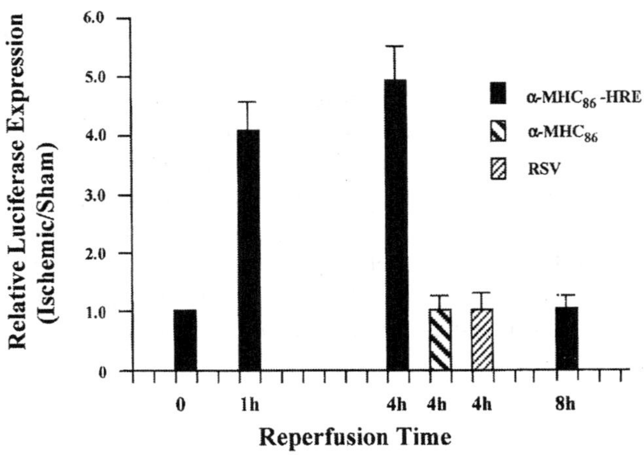

Fig. 10.4. Hypoxia inducible transgene expression in an in vivo model of ischemia with reperfusion. Plasmid DNA was injected into the left ventricle of the rabbit heart. One week later animals were subjected to a 15 minute of myocardial ischemia by ligation of the coronary artery followed by reperfusion. Activity of a hypoxia responsive and muscle specific promoter (HRE/ basal myosin heavy chain gene promoter; alpha-MHC-86-HRE) was assayed by measuring levels of expression of a luciferase reporter gene product. There was rapid activation of the transgene to high levels within one hour reaching a maximum of greater than five fold over four hours and returning to baseline (RSV control levels) by eight hours (after Prentice et al., 1997).

For the administration of neuroprotective agents such as antioxidants or anti-apoptotic factors in ischemic stroke, early administration is considered critical. In these applications the regulated switch has the advantage of responding to hypoxia as an obligatory and integral component of the disease process, thus enabling generation of a therapeutic gene product when the first symptoms of ischemia appear. In the clinical picture of tissue ischemia, where for example multiple ischemic episodes may occur over several hours, the hypoxia regulated gene product would be available for the duration of the ischemia with levels returning to baseline when the ischemic attacks subside.[10,11,68] The identification of directly neuroprotective properties of erythropoietin and VEGF may point to these factors as prime candidates for incorporation into these hypoxia regulated gene transfer vectors in gene therapy for stroke.[59]

But a complicating factor is that the mammalian brain dies in anoxia/ischemia and it is difficult to distinguish between adaptive and pathological events. In this regard the study of the molecular mechanisms that provide protection for the anoxia tolerant species may provide exciting new perspectives. Understanding how these brains regulate channel arrest and are protected from reperfusion free radical generation and the identification of novel anoxia induced genes will provide a resource of key functional targets in ischemic stroke. These targets are likely to be the center of attention not only in many future drug therapies but also for disease appropriate gene therapy interventions based on tissue-specific hypoxia regulated vectors.

4. REFERENCES

1. Abel MS, McCandless DW. Elevated gamma-aminobutyric acid levels attenuate the metabolic response to bilateral ischemia. J Neurochem 1992; 58:740-744.
2. Amoroso S, Schmid-Antomarchi H, Fosset M, Lazdunski M. Glucose, antidiabetic sulfonylureas and neurotransmitter release. Role of ATP-sensitive K^+ channel. Science 1990; 247:852-854.
3. Bickler PE, Gallego SM. Inhibition of brain calcium channels by plasma proteins from anoxic turtles. Am J Physiol 1993; 265:R277-R281.
4. Boening JA, Kass IS, Cottrell JE, Chambers G.. The effect of blocking sodium influx on anoxic damage in the rat hippocampal slice. Neuroscience 1989; 33,:263-268.
5. Boissard CG, Lindner MD, Gribkoff VK. Hypoxia produces cell death in the rat hippocampus in the presence of an A_1 adenosine receptor antagonist: an anatomical and behavioural study. Neurosci 1992; 4:807-812.
6. Busto R, Globus MY, Neary JT, Ginsberg MD. Regional alterations of protein kinase C activity following transient cerebral ischemia: effects of intraischemic brain temperature modulation. J Neurochem 1994; 63:1095-1103.
7. Cordey R, Chiolero R, Miller JA. Resuscitation of neonates by hypothermia: report on 20 cases, with acid-base determination on 10 cases and the long term development of 33 cases. Resuscitation 1973; 2:169-181.
8. Cummings TR, Jiang C, Haddad GG. Human neocortical excitability is decreased during anoxia via sodium channel modulation. J Clin Invest 1993; 91:608-615.

9. Dargent B, Courand F. Down-regulation of voltage-dependent sodium channels initiated by sodium influx in developing neurones. Proc Natl Acad Sci USA 1989; 87:5907-5911.
10. Deanfield JE, Shea M, Ribiero P, de Landsheere CM, Wilson RA, Horlock P, Selwyn AP. Transient ST-segment depression as a marker of myocardial ischemia during daily life. Am J Cardiol. 1984 54(10):1195-200.
11. Deedwania PC, Carbajal EV. Prevalence and patterns of silent myocardial ischemia during daily life in stable angina patients receiving conventional antianginal drug therapy. Am J Cardiol. 1990 65(16):1090-6.
12. Dessi F, Charriaut-Marlangue C, Ben-Ari Y. Anisomycin and cycloheximidide protect cerebellar neurons in culture from anoxia. Brain Res 1992; 581:323-326.
13. Drew KL, Osborne PG, Frerichs KU, Hu Y, Koren RE, Hallenbeck JM, Rice ME., Ascorbate and glutathione regulation in hibernating ground squirrels. Brain Res 1999, 851(1-2):1-8.
14. Fern R, Waxman SG, Ransom BR. Modulation of anoxic injury in CNS white matter by adenosine and interaction between adenosine and GABA. J Neurophysiol 1994; 72:2609-2616.
15. Fern R, Waxman SG, Ransom BR. Endogenous GABA attenuates CNS white matter dysfunction following anoxia. J Neurosci 1995; 15:699-708.
16. Fishman GI, Kaplan ML, Buttrick PM. Tetracycline-regulated cardiac gene expression in vivo. J Clin Invest. 1994, 93(4):1864-8.
17. Frandsen A, Schousboe A. Dantroline prevents glutamate cytotoxicity and Ca^{2+} release from intracellular stores in cultured cerebral neurones. J Neurochem 1991; 56:1075-1078.
18. Frerichs KU Neuroprotective strategies in nature--novel clues for the treatment of stroke and trauma. Acta Neurochir Suppl (Wien) 1999;73(1-2):57-61.
19. Ghersa P, Gobert RP, Sattonnet-Roche P, Richards CA, Merlo Pich E, Hooft van Huijsduijnen R.. Highly controlled gene expression using combinations of a tissue-specific promoter, recombinant adenovirus and a tetracycline-regulatable transcription factor.Gene Ther. 1998 5:1213-20.
20. Ginsberg M, Globus M, Busto R,.The protective effect of mild hypothermia on ischemic brain injury: Role of neurotransmitter release. In: Seylaz J, MacKenzie ET ed(s). Neurotransmission and cerebrovascular function I. Elsevier,1989: 461-464.
21. Ginsberg MD, Sternau LL, Globus M. Theraputic modulation of brain temperature: Relevance to ischemic brain injury. Cerebrovasc. and Brain Metab Revs 1992; 4:189-225.
22. Gracey AY, Troll JV, Somero GN. Hypoxia-induced gene expression profiling in the euryoxic fish Gillichthys mirabilis. Proc Natl Acad Sci U S A. 2001,98:1993-8.
23. Graham SH, Chen J, Sharp FR, et al. Limiting ischemic injury by inhibition of excitatory amino acid release. J Cereb Blood Flow Metab 1993; 13:88-97.
24. Heurteaux C, Lauritzen I, Widmann C, Lazdunski M. Essential role of adenosine, adenosine A1 receptors, and ATP-sensitive K^+ channels in cerebral ischemic preconditioning. Proc Natl Acad Aci USA 1995; 92:4666-4670.
25. Hiramatsu K-I, Kassel NF, Lee KS. Thermal sensitivity of hypoxic responses in neocortical brain slices. J Cereb Blood Flow Metab 1993; 13:395-401.
26. Hochachka PW, Buck LT, Doll CJ ,SC Land.. Unifying theory of hypoxia tolerance : Molecular/ metabolic defense and rescue mechanisms for surviving oxygen lack. Proc. Nat Acad. Sci. USA. 1996, 93, 9493-9498.
27. Hochachka PW and Lutz PL Mechanism, Origin, and Evolution of Anoxia Tolerance in Animals. Comp. Biochem. Physiol.B. 2001, 130, 435-459,
28. Holter JB, Schellens RL. Dantrolene sodium for treatment for carbon monoxide poisoning. Br Med Jn 1988; 296:1772-1773.
29. Ikonomidou C, Mosinger JL, Olney JW. Hypothermia enhances protective effect of MK-801 against hyopxic/ischemic brain damage in infant rats. Brain Res 1989; 487:184-187.
30. Jiang C, Haddad GG. Effect of anoxia on intracellular and extracellular potassium activity in hypoglossal neurons in vitro. J Neurophysiol 1991; 66:103-111.
31. Johansen FF, Diemer NH. Enhancement of GABA neurotransmission after cerebral ischemia in the rat reduces loss of hippocampal CA1 pyramidal neurons. Acta Neurol Scand 1991; 84:1-6.

32. Katsura K-I, Minamisawa H, Ekholm, et al. Changes of labile metabolites during anoxia in moderately hyperthermic rats: correlations to membrane fluxes of K^+. Brain Res 1992; 590:6-12.
33. Kitsis RN, Buttrick PM, McNally EM, Kaplan ML, Leinwand LA. Hormonal modulation of a gene injected into rat heart in vivo.Proc Natl Acad Sci U S A. 1991,88:4138-42.
34. Lekieffre D, Meldrum BS. The pyrimidine-derivative, BW1003C87, protects CA1 and striatal neurons following transient severe forebrain ischemia in rats. A microdialysis study. Neurosci 1993; 56:93-99.
35. Lucas LF, West CA, Rigor BM, et al. Protection against cerebral hypoxia by local anesthetics: a study using brain slices. J Neurosci Meths 1989; 28:47-50.
36. Lyden PD, Lonzo L. Combination therapy protects ischemic brain in rats, a glutamate antagonist plus a gamma-aminobutyric acid agonist. Stroke 1994; 25:189-196.
37. Madden KP. Effect of gamma-aminobutyric acid modulation on neuronal ischemia in rabbits. Stroke 1994; 25:2271-2275.
38. Meldrum BS, Swan JH, Leach MJ, Millan MH, Gwinn R, Kadota K, Graham SH, Chen J, Simon RP.. Reduction of glutamate release and protection against ischemic brain damage by BW 1003C87. Brain Res 1992; 593:106.
39. Minamisawa H, Nordstrom CH, Smith ML, Siesjo BK.. The influence of mild body and brain hypothermia on ischemic brain damage. J Cereb Blood Flow Metab 1990; 10:365-374.
40. Miyashita K, Nakajima T, Ishikawa A, et al. An adenosine uptake blocker, propentofylline, reduces glutamate release in gebril hippocampus following transient forebrain ischemia. Neurochem Res 1992; 17:147-150.
41. Perez-Pinzon M, Rosenthal M, Sick T, Lutz PL. Down-regulation of sodium channels during anoxia: A putative survival strategy of turtle brain. Am J Physiol 1992; 262:R712-R715.
42. Philis JW, Walter GA, Simpson RE. Brain adenosine and transmitter amino acid release from the ischemic rat cerebral cortex: Effects of the adenosine deaminase inhibitor deoxycoformycin. J Neurochem 1991; 56:644-650.
43. Prentice H, Bishopric NH, Hicks MN, Discher DJ, Wu X, Wylie AA, Webster KA. Regulated expression of a foreign gene targeted to the ischaemic myocardium. Cardiovasc Res. 1997 ;35(3):567-74.
44. Ravindran J, Shuaib A, Ijaz S, Galazka P, Waqar T, Ishaqzay R, Miyashita H, Liu L. High extracellular GABA levels in hippocampus - as a mechanism of neuronal protection in cerebral ischemia in adrenalectomized gerbils. Neurosci Lett 1994; 176:209-211.
45. Rice ME, Cammack J. Anoxia-resistant turtle brain maintains ascorbic acid contents in vitro. Neurosci Lett 1991; 132:141-145.
46. Robbins PD, Ghivizzani SC. Viral vectors for gene therapy. Pharmacol Ther. 1998 ;80:35-47..
47. Rothman SM. Excitotoxic neuronal death: mechanisms and clinical relevance. Seminars in The Neurosciences 1994; 6:315-322.
48. Rothman SM, Olney JW. Excitotoxicity and the NMDA receptor - still lethal after eight years. Trends Neurosci 1995; 18:57-58.
49. Sancibrian M, Serrano JS, Minano FJ. Opioid and prostaglandin mechanisms involved in the effects of GABAergic drugs on body temperature. Gen Pharmacol 1991; 22:259-262.
50. Schaeffer P, Lazdunski M. K^+ efflux pathways and neurotransmitter release associated to hippocampal ischemia: effects of glucose and of K^+ channel blockers. Brain Res 1991; 539:155-158.
51. Schwartz RD, Huff RA, Yu X. Postischemic diazepam is neuroprotective in the gerbil hippocampus. Brain Res 1994; 647:153-160.
52. Schwartz RD, Yu X, Katzman MR, Hayden-Hixson DM, Perry JM.. Diazepam, given postischemia, protects selectively vulnerable neurons in the rat hippocampus and striatum. J Neurosci 1995; 15:529-539.
53. Shuaib A, Ijaz S, Hasan S, Kalra J.. Gamma-vinyl GABA prevents hippocampal and substantia nigra reticulata damage in repetitive transient forebrain ischemia. Brain Res 1992; 590:13-17.
54. Shuaib A, Mazagri R, Ijaz S. GABA agonist "muscimol" is neuroprotective in repetitive forebrain ischemia in gerbils. Exp Neurol 1993; 123:284-288.
55. Siebke H, Rod T, Breivik H, Link B.. Survival after 40 minutes submersion without cerebral sequelae. Lancet 1975; 7919:1275-1277.

56. Siesjo BK, Smith ML. Brain resuscitation: Yesterday , Today and Tomorrow. In: Takeshita H, Siesjo BK, Miller JD ed(s). Advances in Brain Resuscitation. Berlin: Springer,1988: 3-19.
57. Siesjo BK. Pathophysiology and treatment of focal cerebral ischemia. Part II: Mechanisms of damage and treatment. J Neurosurg 1992; 77:337-354.
58. Silver B, Weber J, Fisher M. Medical therapy for ischemic stroke. Clin Neuropharmacol 1996; 19:101-128.
59. Sondell M, Lundborg G, Kanje M. Vascular endothelial growth factor has neurotrophic activity and stimulates axonal outgrowth, enhancing cell survival and Schwann cell proliferation in the peripheral nervous system. J Neurosci. 1999 ;19(14):5731-40.
60. Stea A, Jackson A, Nurse CA. Hypoxia and N^6, $O^{2'}$-dibutyryladenosine 3',5'-cyclic monophosphate, but not nerve growth factor, induce Na^+ channels and hypertrophy in chromaffin-like arterial chemoreceptors. Proc Natl Acad Sci 1992; 89:9496-9473.
61. Steen PA, Soule EH, Michenfelder JD. Detrimental effects of prolonged hypothermia in cats and monkeys with and without regional cerebral ischemia. Stroke 1979; 10:522-533.
62. Steen PA. The role of calcium entry blockers in brain ischemia. In: Takeshita H, Siesjo BK, Miller JD ed(s). Advances in Brain Resuscitation. Berlin: Springer,1988: 211-219.
63. Swanson RA, Chen J, Graham SH. Glucose can fuel glutamate uptake in ischemic brain. J Cereb Blood Flow and Metab 1994; 14:1-6.
64. Sweeney MI. Neuroprotective effects of adenosine in cerebral ischemia: window of opportunity. Neurosci and Biobehav Revs 1997 20 in press.
65. Urenjak J, Obrenovitch TP. Pharmacological modulation of volyage-gated Na^+ channels: A rational and effective strategy against ischemic brain damage Pharmacol Rev 1996; 48: 21-67.
66. Valentino K, Newcomb R, Gadbois T, Singh T, Bowersox S, Bitner S, Justice A, Yamashiro D, Hoffman BB, Ciaranello R. A selective N-type calcium channel antagonist protects against neuronal loss after global ischemia. Proc Natl Acad Sci USA 1993; 90:7894-7897.
67. Vannucci RC, Mujsce DJ. Effect of glucose on perinatal hypoxic-ischemic brain damage. Biol Neonate 1992; 62:215-224.
68. Webster, KA. Molecular switches for regulating therapeutic genes. Gene Therapy 1998, 6, 951-953
69. Wenger RH. Mammalian oxygen sensing, signalling and gene regulation. J Exp Biol. 2000;203 :1253-63.
70. Ye X, Al-Babili S, Kloti A, Zhang J, Lucca P, Beyer P, Potrykus I. Engineering the provitamin A (beta-carotene) biosynthetic pathway into (carotenoid-free) rice endosperm. Science. 2000 ;287(5451):303-5.
71. Xia Y, Haddad GG. Major differences in CNS sulfonylurea receptor distribution between the rat (newborn, adult) and turtle. J Comp Neur 1991; 314:278-289.
72. Xie Y, Dengler K, Zacharias E, et al. Effects of sodium channel blocker tetrodotoxin (TTX) on cellular ion homeostasis in rat brain subjected to complete ischemia. Brain Res 1994: 652:216-224.

INDEX

E

F

G

H